AF678101

Hybrid O/I Sol–Gel-Derived Nanocomposites Systems for Advanced Functional Applications

Hybrid O/I Sol–Gel-Derived Nanocomposites Systems for Advanced Functional Applications

Guest Editors

Giulio Malucelli
Aurelio Bifulco

MDPI

Basel • Beijing • Wuhan • Barcelona • Belgrade • Novi Sad • Cluj • Manchester

Guest Editors
Giulio Malucelli
Department of Applied Science and Technology
Politecnico di Torino
Torino
Italy

Aurelio Bifulco
Department of Chemical, Materials and Production Engineering (DICMaPI)
University of Naples Federico II
Naples
Italy

Editorial Office
MDPI AG
Grosspeteranlage 5
4052 Basel, Switzerland

This is a reprint of the Special Issue, published open access by the journal *Molecules* (ISSN 1420-3049), freely accessible at: https://www.mdpi.com/journal/molecules/special_issues/7GG0GHJ414.

For citation purposes, cite each article independently as indicated on the article page online and as indicated below:

Lastname, A.A.; Lastname, B.B. Article Title. *Journal Name* **Year**, *Volume Number*, Page Range.

ISBN 978-3-7258-5663-3 (Hbk)
ISBN 978-3-7258-5664-0 (PDF)
https://doi.org/10.3390/books978-3-7258-5664-0

Cover image courtesy of Aurelio Bifulco

© 2025 by the authors. Articles in this book are Open Access and distributed under the Creative Commons Attribution (CC BY) license. The book as a whole is distributed by MDPI under the terms and conditions of the Creative Commons Attribution-NonCommercial-NoDerivs (CC BY-NC-ND) license (https://creativecommons.org/licenses/by-nc-nd/4.0/).

Contents

About the Editors

Giulio Malucelli

Giulio Malucelli is a Professor of materials science and technology in the Department of Applied Science and Technology at Politecnico di Torino in Italy. He holds a Ph.D. in chemistry. He heads the "ALL-Polymer" group at the Alessandria branch of the university. According to the Scopus database, as of September 2025, he has co-authored more than 330 peer-reviewed articles and received over 12,900 total citations from 7,400 documents; his h-index is 66. Giulio Malucelli has also published 60 book chapters and one book and contributed over 290 communications to national and international congresses. In 2022, 2023, and 2024, Malucelli was included in Stanford's "Top 2% Scientists" list (Updated Science-Wide Author Databases of Standardized Citation Indicators, Elsevier Data Repository). His current research activities include sol–gel-derived systems, flame retardancy in polymers and their micro- to nano-composites and investigating structure–property–processing relationships in polymer systems.

Aurelio Bifulco

Aurelio Bifulco is an Adjunct Professor of Chemistry at the University of Naples Federico II (UniNa). Before that, he was an Assistant Professor in the Department of Chemical, Materials and Production Engineering at UniNa. His research activity is focused on sol–gel chemistry, polymer chemistry, flame retardancy, and nanotechnology. In 2025, Aurelio Bifulco was a Visiting Professor at the University of Lorraine, France. In 2023 and 2025, he was an invited speaker at the European Meeting on Fire Retardant Polymeric Materials (FRPM). In 2020, he received his PhD in Industrial Product and Process Engineering from UniNa. In 2019, he was a Visiting PhD Student at the Swiss Federal Laboratories for Materials Science and Technology (Empa), where he worked as a member of the Additive and Chemistry Group on the development of new hybrid flame-retardant strategies for epoxy materials. In 2014, as an Erasmus student, he carried out a master's degree thesis in the Institute of Chemical Engineering at TU Wien, conducting research on combustion and fluidized bed systems. In 2011, he carried out a bachelor's degree thesis at UniNa, studying the synthesis of liquid crystalline polymers for optical applications. Aurelio Bifulco is a Reviewer for international grant proposals from the National Science Centre of Poland, the Hungarian Academy of Sciences, and the European Commission. He serves as an Editorial Board Member for the *Journal of Vinyl and Additive Technology* (Wiley) and as an Associate Editor for *Polymers from Renewable Resources* (Sage Journals). Finally, he is the owner and inventor of two Italian patents, and has published 3 book chapters and more than 54 articles in international peer-reviewed journals that attracted 1,321 citations; his h-index is 22.

Preface

The need for high-performing polymer-based systems has continued to propel (and is still encouraging) both the academic and industrial communities toward the design, synthesis, and characterization of novel, advanced, and sustainable materials that are suitable for application in demanding sectors (i.e., electrical engineering and electronics, advanced packaging, and flame retardance, among others). This Reprint collects significant contributions from scholars working in the field of polymer research and in the development of hybrid O/I sol–gel-derived nanocomposite systems for advanced functional applications. It reports the recent progress and trends in the preparation, characterization, applications, processability, and sustainability of hybrid O/I systems. This Reprint provides in-depth insights into the synthesis of hybrid O/I sol–gel-derived nanocomposites and their growing applications. It describes the interactions between polymer matrices and functional components, emphasizing recent advancements in material design and performance. The volume also explores alternative fillers with environmental and economic benefits and highlights the increasing role of machine learning in sol–gel science. These data-driven approaches enhance the prediction of material properties, the optimization of formulations, and the discovery of new composites. Each chapter provides readers with a deeper understanding of current developments and the impact of artificial intelligence in materials science by offering a solid theoretical background, experimental results, and key conclusions.

Giulio Malucelli and Aurelio Bifulco
Guest Editors

molecules

MDPI

Article

Bridged Mesoporous Oxo-Phosphonates: A General Strategy Toward Functional, Hybrid Materials

Elodie Gioan [1], Zijie Su [1], Yanhui Wang [2], Jeremy Rodriguez [1], Karim Bouchmella [1] and Johan G. Alauzun [1,*]

1 ICGM, University of Montpellier, CNRS, ENSCM, 34293 Montpellier, France; elodie.gioan@umontpellier.fr (E.G.); karim.bouchmella@umontpellier.fr (K.B.)
2 Yantai Research Institute, Harbin Engineering University, 1 Qingdao Street, Development Zone, Yantai 150009, China
* Correspondence: johan.alauzun@umontpellier.fr

Abstract: Combining the properties of organic and inorganic components with high surface areas and large pore volumes opens up countless possibilities for designing materials tailored to a wide range of advanced applications. As the majority of mesoporous hybrid materials are siliceous, the development of cost-effective synthetic approaches to produce water-stable hybrids with controlled porosity and functionality remains essential. Herein, we describe an original strategy for the synthesis of bridged mesoporous titania–bisphosphonate hybrids based on a one-step, template-free, non-hydrolytic sol–gel process. The reaction between $Ti(OiPr)_4$ and several flexible or rigid bisphosphonate esters, in the presence of acetic anhydride (Ac_2O) leads to the formation of TiO_2 anatase nanorods interconnected by fully condensed bisphosphonate groups. The general method that we depict is quantitative and low cost. All materials are mesoporous with very high specific surface areas (up to 520 $m^2 \cdot g^{-1}$) and pore volumes (up to 0.93 $cm^3 \cdot g^{-1}$).

Keywords: hybrids; bisphosphonate; titania; mesoporosity

1. Introduction

The synthesis of porous organic–inorganic hybrid materials has been a major challenge for materials scientists for more than 30 years [1,2]. Indeed, these materials tend to combine the properties of organic and inorganic moieties with high surface areas and pore volumes. They should offer unprecedented options to design many materials adapted to a huge range of advanced applications. These hybrid materials are often separated into two different classes. In Class I hybrid materials, the organic and inorganic parts are bounded through weak bonds, whereas, in Class II hybrid materials, they are bounded toward stronger ionocovalent or covalent bonds.

Most Class II hybrid materials are siliceous derivatives, typically derived from organotrialkoxysilane coupling agents. These compounds have been extensively employed to develop a wide variety of hybrid materials exhibiting either disordered or ordered mesoporosity, including surface-functionalized silicas, (bridged) silsesquioxane gels and cogels, and periodic mesoporous organosilicas [3–6].

Research is increasingly focusing on nonsiliceous mesoporous hybrid materials to extend the applications. They are mainly developed using carboxylate (MOFs) [7,8] or phosphonate coupling agents [9–13].

On the past two decades, metal-organic frameworks (MOFs) have known an exponential growth [14,15]. They are notorious for their nice crystalline structures, extremely large surface areas, and well-defined micropores. However, designing mesoporous MOFs often

 https://doi.org/10.3390/molecules30112459

requires elaborate strategies, and the limited water stability of MOFs might be an issue, depending on the applications [16].

On the other hand, phosphonates, known for their strong binding affinity to a broad range of metal atoms, are considered among the most promising coupling agents for designing water-stable, nonsiliceous Class II hybrid materials [17,18]. However, traditional metal phosphonates typically form dense, layered structures with inherently low porosity [19]. Various strategies have been employed to synthesize porous metal phosphonates based on titanium, zirconium, or aluminum [9,20–23]. One common approach involves the combination of bridging bisphosphonates with short monophosphonates to induce microporosity. Another method utilizes bulky multiphosphonic acids to prevent the formation of layered structures [24]. Additionally, metal phosphonates exhibiting external mesoporosity have been developed through sol–gel and/or hydrothermal techniques using bisphosphonic or multiphosphonic acids [19,25]. In these systems, the structural walls are composed of amorphous or semicrystalline metal phosphonate particles, characterized by high phosphorus-to-metal (P/M) ratios, typically ranging from about 1.3 to 2 [26–28]. However, the variety of functional organic groups incorporated into metal phosphonate hybrid materials remains relatively limited [29].

Therefore, establishing a template-free and versatile method for synthesizing nonsiliceous hybrids with controllable mesoporosity and customizable functionality remains a highly sought-after goal.

We have lately reported the non-hydrolytic sol–gel (NHSG) synthesis of a new family of hybrid organic-inorganic mesoporous materials with tunable porosity and functionality [30,31]. We called them "bridged mesoporous oxo-phosphonates", or BMOPs. In this previous work, the hybrids consisted of titanium oxide nanodomains interconnected by rigid, fully condensed bisphosphonate linkers (biphenyl or bipyridine). Their porosities were shown to be easily tunable via reaction conditions, and they exhibited excellent stability across a wide pH range. We believe that the NHSG chemistry provides an elegant, simple, and powerful route to obtain hybrids with different morphologies or textures.

Herein, we report the one-pot synthesis of several hybrids containing different bisphosphonates (alkyl, aryl, and bipyridine derivatives) with TiO_2. As we describe, this simple method, giving access to a wide family of new mesoporous hybrid materials, is generic and can be extended to different flexible or rigid bisphosphonate groups. All materials show specific surface areas between 290 and 520 $m^2\ g^{-1}$, as well as pore volumes up to 0.93 $cm^3\ g^{-1}$. We consider these materials to be promising as catalytic supports or sorbents given their high stability and outstanding textural properties.

2. Results and Discussion

In our previous work, we demonstrated that TiO_2 with hierarchical porosity and high specific surface area can be synthesized through an NHSG reaction between $Ti(OiPr)_4$ and Ac_2O [30]. We further showed that coupling this NHSG approach with a rigid bis-diethylphosphonate precursors provides a one-step and template-free method for the synthesis of nonsiliceous hybrid materials with tunable mesoporosity and customizable post-synthetic functionalization.

The synthesis of these bridged mesoporous oxo-phosphonates has been shown to be a three-step process (Equations (1)–(3) in Figure 1), where two are competitive reactions (Equations (2) and (3)). In the first step, titanium isopropoxide reacts with acetic anhydride, forming Ti-OAc intermediates (Equation (1)). These functions then react either with other titanium isopropoxide functions (Equation (2)) or with phosphonate esters (Equation (3)), where R is any functional organic group.

$$\text{Ti-O}^i\text{Pr} + (\text{CH}_3\text{CO})_2\text{O} \longrightarrow \text{Ti-OCOCH}_3 + \text{CH}_3\text{COO}^i\text{Pr} \quad (1)$$

$$\text{Ti-O}^i\text{Pr} + \text{Ti-OCOCH}_3 \longrightarrow \text{Ti-O-Ti} + \text{CH}_3\text{COO}^i\text{Pr} \quad (2)$$

$$\text{RP-OEt} + \text{Ti-OCOCH}_3 \longrightarrow \text{RP-O-Ti} + \text{CH}_3\text{COOEt} \quad (3)$$

Figure 1. Reaction involved in the BMOP synthesis.

Being irreversible, TiO_2 grows and, in time, is modified on the surface with phosphonate moieties. These simultaneous reactions form BMOP hybrids (Scheme 1).

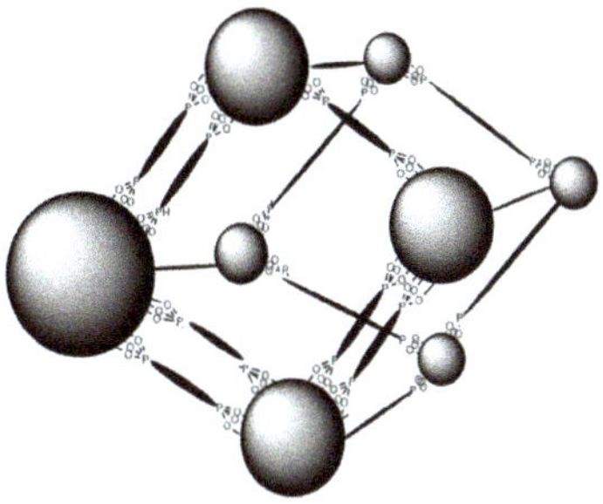

Scheme 1. Schematic representation of BMOP (metal oxide nanoparticles bridged by bisphosphonate).

A major advantage of hybrid materials is the possibility to tailor the functional organic groups to the targeted applications. In order to give access to a wide family of these new mesoporous hybrid materials with great application potential, we have considered here to keep the titania precursor and to change the organic linker while maintaining the P/Ti ratio (0.1 eq). Seven bisphosphonate precursors were studied (Figure 2). Three are flexibles: propyl (bC_3P), hexyl (bC_6P) and dodecyl ($bC_{12}P$); three are rigids: phenyl (bPhP), biphenyl (bPh_2P) and bipyridine (bPy_2P); and one is semi-rigid: dimethylbipyrine (bPy_2C_2P). It is worth noting that all these molecules are commercially available but are also synthesizable.

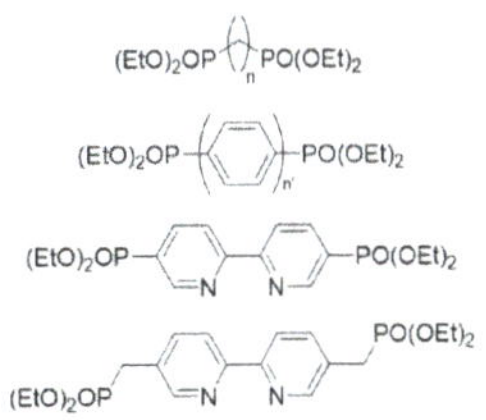

Figure 2. All bisphosphonates considered (n = 3, 6, and 12; n' = 1 and 2).

Using a previously described procedure [30] (as depicted in Scheme 2 and developed in the experimental section), we thus obtained 7 different BMOPs: bC_3P-Ti, bC_6P-Ti, $bC_{12}P$-Ti, bPhP-Ti, bPh_2P-Ti, bPy_2P-Ti, and bPy_2C_2P-Ti.

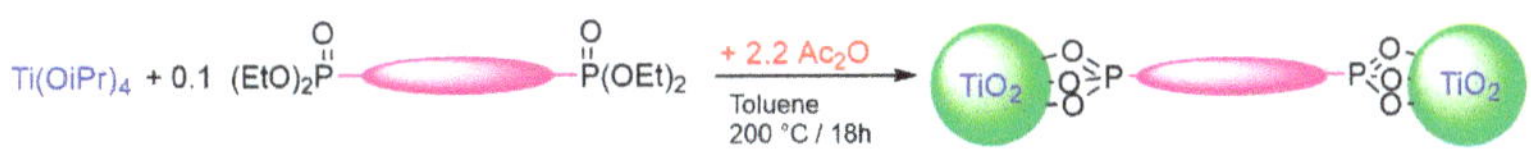

Scheme 2. Schematic representation of BMOP syntheses.

For all obtained hybrids, we measured the P/Ti ratios by MEB–EDX. According to the results, all the precursors are incorporated and the materials are homogeneous at the micrometer scale (P/Ti = 0.19 ± 0.02), within experimental error. It is worth noting that the synthetic yields are always quantitative. The nitrogen adsorption–desorption isotherms of the different samples are displayed in Figure 3a (flexible linkers) and 3b (rigid linkers and bPy_2C_2).

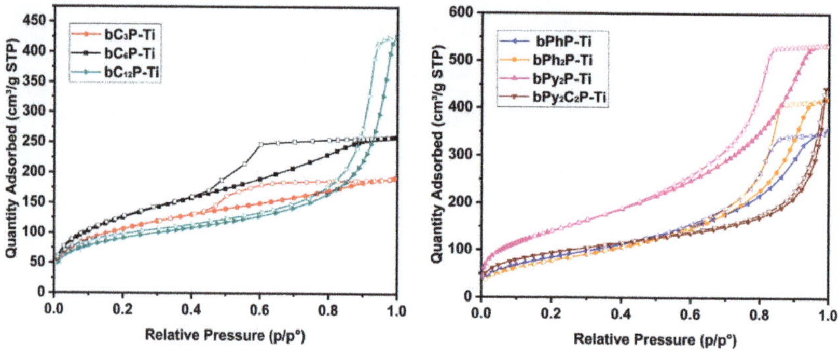

(a) (b)

Figure 3. Nitrogen physisorption isotherms of flexible (**a**) and rigid (**b**) BMOP samples. Filled and open symbols refer to adsorption and desorption, respectively.

As we can observe, all samples showed significant porosity. In accordance with the IUPAC classification [32], the isotherms of theses BMOP samples are mainly of type IVa, characteristic of mesoporous adsorbents, with an H2 hysteresis loop indicating complex pore structures. The isotherms of bPy_2C_2 also showed Type II features (lack of plateau at high relative pressure), suggesting the presence of some macropores.

As shown in Table 1, the specific surface areas of the flexible bisphosphonates are between 330 and 470 $m^2\ g^{-1}$. Indeed, for $bC_{12}P$-Ti, the specific surface area is 330 $m^2\ g^{-1}$; it is 470 $m^2\ g^{-1}$ for bC_6P-Ti, and it reaches 390 $m^2\ g^{-1}$ for bC_3P-Ti (Table 1). There is, thus, an optimal value for the bC_6P bisphosphonate linker. A clear trend can be observed: the pore volume decreases progressively with decreasing chain length—0.30 (bC_3P-Ti), 0.40 (bC_6P-Ti), and 0.60 $cm^3\ g^{-1}$ ($bC_{12}P$-Ti). Increasing the distance between nanoparticles should indeed increase interparticle volumes.

Table 1. Recapitulative data from N_2 sorption, XRD, and TGA measurements.

Sample	S_{BET} [a] ($m^2\ g^{-1}$)	V_p [b] ($cm^3\ g^{-1}$)	V_m [c] ($cm^3\ g^{-1}$)	D_p [d] (nm) BJH	D_p [e] (nm)	C_{BET} [f]	d_{cryst} [g] (nm)	Crosslinking Rate (%)
bC_3P-Ti	390	0.30	<0.01	4.0	4.0	72	14.6	96
bC_6P-Ti	470	0.40	<0.01	2.7	4.0	56	11.7	89
$bC_{12}P$-Ti	330	0.66	0.09	5.4	25	134	amorphous	89
bPhP-Ti	315	0.54	<0.01	9.6	13	54	12.8	96
bPh_2P-Ti	290	0.65	<0.01	7.1	13	47	8.5	96
bPy_2P-Ti	520	0.93	0.06	6.5	10	46	8.9	86
bPy_2C_2P-Ti	345	0.69	0.05	10.9	-	79	10.1	97

[a] BET specific area. [b] Total pore volume at $P/P_0 = 0.99$. [c] Volume of micropores estimated by DFT analysis. [d] BJH average pore diameter in the 2 to 50 nm range from the desorption branch. [e] Pore diameter in the 2 to 50 nm range of the desorption branch. [f] Brunauer–Emmett–Teller C constant of $0.05 < P/P_0 < 0.3$. [g] Crystallite size estimated by the Scherrer equation.

The specific surface areas of the BMOPs containing rigid bisphosphonates are in the range of 290 $m^2\ g^{-1}$ to 520 $m^2\ g^{-1}$. It is somewhat more challenging to compare these data as the functions are different. Nevertheless, a decrease in specific surface area is observed from bPhP to bPh_2P, while the pore volume shows a concomitant increase.

The pore size distribution (Figure 4a,b) results confirmed the presence of mesopores in all BMOP samples. As depicted in these figures and summarized in Table 1, the pore

diameters are centered within the range of 4.0 to 25 nm for materials with flexible linkers and around 10–13 nm for the ones with rigid linkers. The average pore diameter (estimated with BJH) is within 2.7 to 5.4 for the flexible linkers and within 6.5 to 10.9 nm for the rigid ones. As observed, the semi-rigid linker does not show evidence of porous diameters in the mesoscopic range. The porosity might be higher than 50 nm.

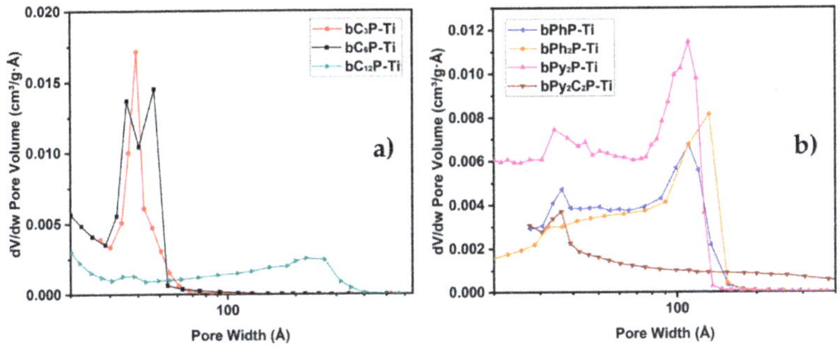

Figure 4. Pore sizes distribution for flexible (**a**) and rigid (**b**) BMOP samples.

The peak found in the distribution at ≈4.0 nm for some samples (such as bC_6P-Ti) is a well-known artefact. This merely indicates the presence of small pores of diameters < 4 nm, due to the instability of the meniscus at relative pressures lower than 0.42 [33].

We finally determined the Brunauer–Emmett–Teller C constant of $0.05 < P/P_0 < 0.3$, which is related to the adsorption enthalpy, for all BMOPs. The C constant represents the affinity of the solid with the adsorbate (the N_2 molecules) and also to the heat of adsorption. The higher the value of C, the higher the interaction [32,34]. The highest C values were recorded for the flexible linkers, although all values remain within the 46–134 range.

Relatively similar C constant values (≈30) were previously reported for nanoparticles grafted by octylphosphonic acid [35].

TiO_2 prepared by the exact same non-hydrolytic procedure in toluene, is purely mesoporous, with a pore volume of 0.36 cm^3 g^{-1} and a specific surface area of 170 m^2 g^{-1} [36]. As previously described, the presence of bridging bisphosphonate groups leads to the formation of a high specific surface area. Thus, this general method allows one to obtain highly mesoporous material with all studied bisphosphonates.

The ^{13}C MAS NMR (Figure S1) performed on $bC_{12}P$-Ti (a) and bPy_2P-Ti (b) samples confirmed the incorporation of the organic linkers without alteration during the synthetic process. As observed in the spectra, the absence of resonances at ≈60 ppm (OCH_2CH_3 sites) and ≈75 pmm ($OCH(CH_3)_2$ sites) reveals the full condensation of P-OEt and Ti(OiPr) groups. We can notice the presence of two signals centered at 23 ppm and 178 ppm, pointing to the presence of residual acetate groups within the materials. ^{31}P MAS solid-state NMR spectrum (Figure 5) of the bC_6P-Ti hybrid material displays a very broad signal in the 10 to 40 ppm range, centered at 27 ppm.

We have reported similar broad resonances for TiO_2–octylphosphonate hybrid materials prepared in the same one-step NHSG process, starting from $Ti(OiPr)_4$ and H_3C-$(CH_2)_7$-$PhPO(OEt)_2$ [35]. These peaks confirm the presence of phosphonate species bounded to the TiO_2 particles through Ti–O–P bonds. The lack of a sharp resonance around 7 ppm demonstrates the absence of a layered titanium bisphosphonate phase. The deconvolution determined by OriginPro software (2022b SR1; v9.9.5.171) shows two contributions at 27 ppm (85%) and 17.3 ppm (15%). As previously demonstrated. this is consistent with the tridentate bonding of R-PO_3 onto the titania surfaces [35].

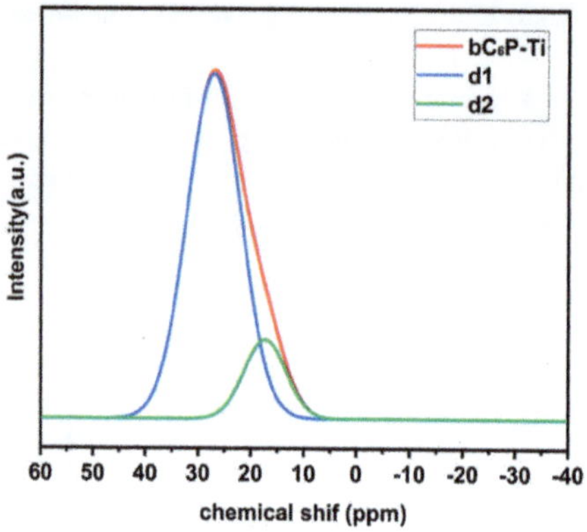

Figure 5. ^{31}P CP magic angle spinning solid-state NMR spectrum of the bC_6P-Ti hybrid materials. The d1 and d2 curves represent the deconvolution of the bC_6P-Ti curve.

Electron microscopy images (Figure S2) show that the bPh_2P-Ti material appeared to be formed of densely aggregated, roughly spherical particles.

As shown in the TEM images (Figure 6), at lower scale, the particles form a three-dimensional mesoporous network, demonstrating the generality of our synthetic route. Interestingly, BMOPs obtained from a rigid linker appears to be constituted of TiO_2 nanorods (Figure 6c,d), while the ones prepared from flexibles linker (bC_6P-Ti and $bC_{12}P$-Ti) exhibit an assembly of spherical nanoparticles (10 ± 5 nm diameter). The nanorods have a length of 60 ± 15 nm and 10 ± 2 nm width. The morphology of the samples is thus significantly dependent on the organic linker. At this nanometer scale, the sample textures are thus significantly dependent on the rigidity of the organic linker. We do not have at this time an explanation for these interesting differences.

The X-ray diffraction (XRD) patterns of the hybrid samples and TiO_2 are presented in Figure 7.

The samples' crystallinities do not differ much, except for the $bC_{12}P$-Ti sample, which exhibits amorphous structure. The patterns of the BMOP hybrids showed the presence of anatase nanocrystals (JCPDS 21-1272). There was no evidence of any rutile phase. In all cases and as previously described, low-angle XRD showed no evidence for layered titanium bisphosphonate phase. Given that this stable phase is generally non-porous, its formation could adversely affects organic group accessibility and compromises the material's textural characteristics. The sizes of the anatase crystallites determined by the Debye–Scherrer equation from the (101) and (200) peaks of the powder X-ray diffraction patterns vary from 8.5 to 14.6 nm. As observed, TiO_2 nanomaterials synthesized with the exact same procedure have the same anatase phase.

The attenuated total reflection (ATR)—Fourier transform infrared (FTIR) spectra of hybrid materials display a single broad band between 1000 and 1100 cm^{-1}, attributed to the vibrational modes of R-PO_3 tetrahedra (Figure S3). The absence of characteristic bands at $\approx$1220 cm^{-1} (P=O stretching vibration) and $\approx$950 cm^{-1} (P–OC stretching vibrations) [37] indicates that the phosphonate groups adopt a tridentate coordination environment, similar to that found in layered titanium phosphonates. As previously observed in sol–gel-derived TiO_2–phenylphosphonate hybrid materials, P atoms are bonded to three Ti atoms in $CP(OTi)_3$ sites [38]. Vibrations observed in the 1400–1500 cm^{-1} region are assigned to CH_3 and CH_2 deformations associated with residual organic groups on the surface (e.g., Ti–OiPr, Ti–O–CMePhOiPr), as well as CH_3, CH_2, and P–CH_2 deformations originating from the alkylphosphonate linkers (C_3, C_6, and C_{12}). The three bands between

2850 and 3000 cm^{-1} are ascribed to the C–H sp3 symmetric and asymmetric stretching vibrations of bonds in CH_2 and CH_3, mostly in the dodecyl unit. The intensity of these bands is directly related to the amount of Csp_3.

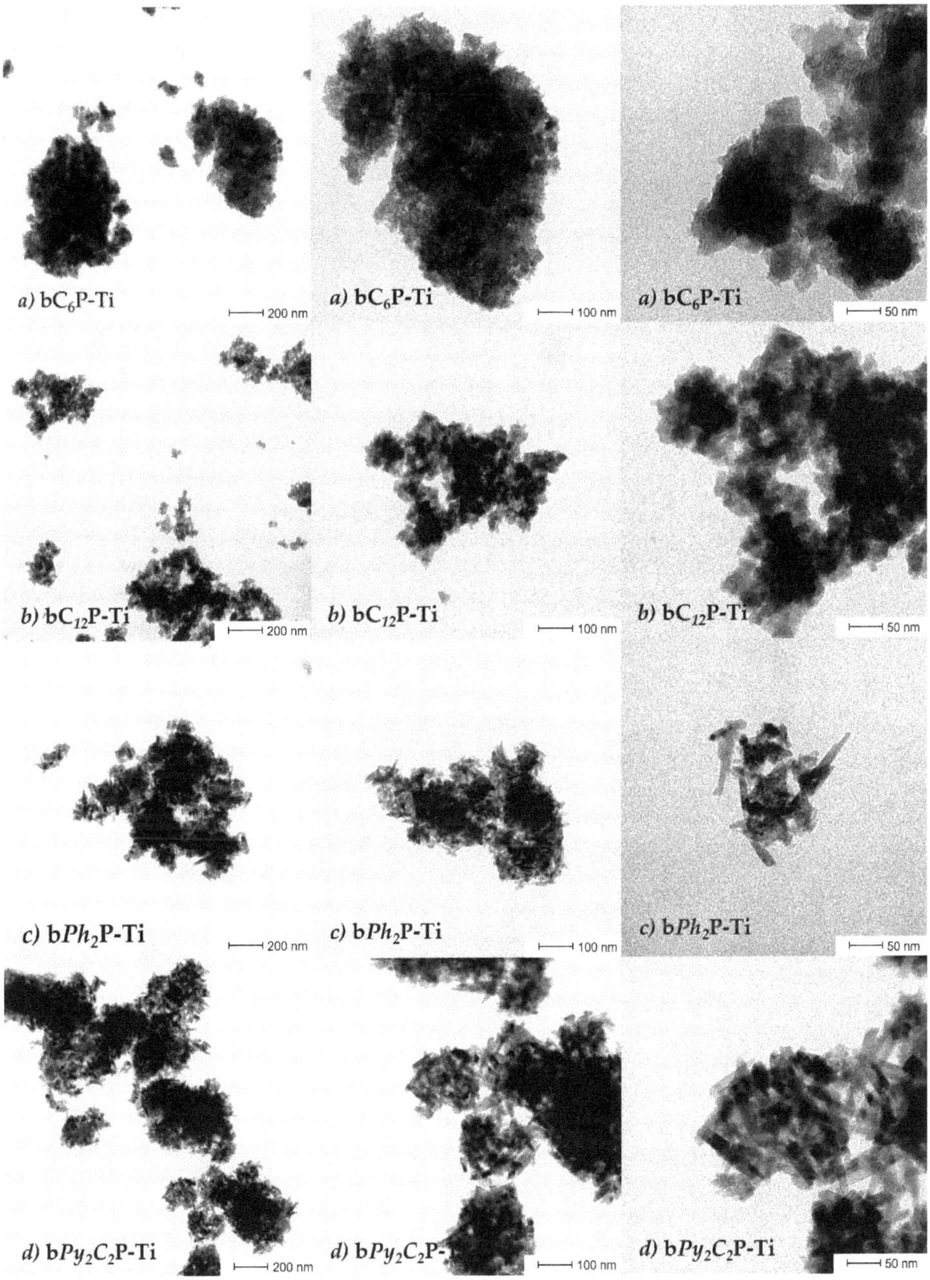

Figure 6. TEM images with scale bars at 50, 100 and 200 nm; of bC_6P-Ti (**a**), $bC_{12}P$-Ti (**b**), bPh_2P-Ti (**c**), and bPy_2C_2P-Ti (**d**) samples.

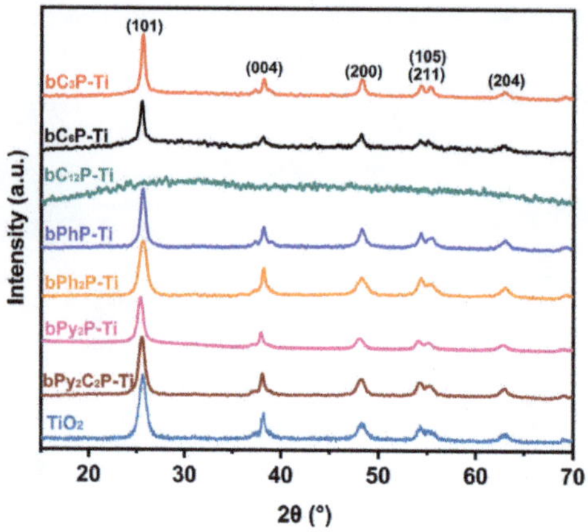

Figure 7. Powder X-ray diffraction (XRD) patterns of BMOP and TiO_2 samples.

A weak, broad band between 3000 and 3800 cm^{-1}, characteristic of O–H stretching vibrations, indicates the presence of a small amount of adsorbed water, corroborated by the band at 1620 cm^{-1}, corresponding to its bending mode. This broad feature may also originate from hydroxyl groups generated by the hydrolysis of residual surface species during the washing process or manipulation under air.

In order to confirm the incorporation of bisphosphonate units and to determine the degree of condensation, we undertook thermogravimetric measurements of all BMOP materials from room temperature up to 650 °C. We rescaled all data curves to 0% mass lost at 150 °C. According to TGA (Figure S4 for flexible BMOPs and Figure S9 for rigid BMOPs), all BMOPs gradually lose mass upon increasing the temperature up to 650 °C. This loss is mainly due to the decomposition related to carbon and hydrogen oxidation, as well as the oxidation of nitrogen, resulting in the release of nitrogen oxides (NO_x). The major decomposition range is within the range of 300–500 °C. This demonstrates that these materials do not present higher stability under air than siliceous hybrids materials.

The lost mass is in good agreement with the total number of organic units inserted into the hybrids. The values are between 12 and 31% for bC_3P-Ti and $bC_{12}P$-Ti. The crosslinking rate or degree of condensation lies between 82 and 96%. This result is consistent with the ^{13}C data (no remaining ethoxy groups and few Ti-Ac).

In order to study the stability of these hybrids, we immersed the $bC_{12}P$-Ti material for 24 h at room temperature at pH 3 and at pH 11 under stirring. After further rinsing and drying steps (same conditions as before), the materials were characterized by nitrogen sorption. The isotherms are presented in the "additional information" section (Figure S5). As can be seen, there is no change in the mesoporosity part of the isotherm (type IV, H2). There is, however, a reduction in the microporosity section and, therefore, a significant decrease in specific surface area (approx. 50%). Indeed, the surface area drops from 330 $m^2\ g^{-1}$ to 168 $m^2\ g^{-1}$ (pH 3) and 162 $m^2\ g^{-1}$ (pH 11). Pore volumes are reduced by 20% (from 0.66 $cm^3\ g^{-1}$ to 0.51 $cm^3\ g^{-1}$ and 0.50 $cm^3\ g^{-1}$, respectively). Hybrids obtained with flexible precursors therefore appear to be slightly less stable than those obtained with rigid precursors [30].

3. Materials and Methods

Titanium (IV) isopropoxide ($Ti(OiPr)_4$, 97%) and acetic anhydride (Ac_2O, 99%) were purchased from Sigma-Aldrich (St. Quentin Fallavier, France). Tetraethyl 1,3 propyl-bisphosphonate (bC_3P, 95%), tetraethyl 1,6 hexyl-bisphosphonate (bC_6P, 95%), tetraethyl 1,12 dodecyl-bisphosphonate ($bC_{12}P$, 95%), tetraethyl 1,4 phenyl-bisphosphonate (bPhP, 97%), tetraethyl 1,1′-biphenyl-4,4′-bisphosphonate (bPh_2P, 97%), tetraethyl 2,2′-bipyridine-5,5′-bisphosphonate (bPy_2P, 97%), and tetraethyl 2,2′-bipyridine-5,5′-dimethylbisphosphonate (bPy_2C_2P, 95%) were purchased from SiKEMIA (Montpellier, France).

All reactants were used without further purification. Toluene (Sigma-Aldrich 99.7%) was obtained over a Pure Solve MD5 solvent purification system (H_2O < 10 ppm) and controlled with a Karl Fischer coulometer.

3.1. Synthesis of Hybrid Materials

To avoid the presence of water, the addition of reactants and the sealing of the autoclaves were carried out in a glovebox under an argon atmosphere (<10 ppm of H_2O and O_2).

In a typical synthesis, titanium (IV) isopropoxide (1.50 g, 5.25 mmol) was solubilized in toluene (8 mL), before adding the bisphosphonate (0.53 mmol). After 10 min of stirring, the acetic anhydride (2.2 eq. 1.18 g, 11.55 mmol) was added slowly. After stirring for another 10 min, the obtained solution was transferred to a stainless-steel autoclave (Parr Instruments) with a PTFE lining (23 mL), which was then sealed. The autoclave (fill rate of approx. 45%) was heated in an oven at 200 °C under autogenous pressure for 18 h. After reaction, the resulting monolithic gel was thoroughly washed with hot ethanol (Soxhlet for 24 h) and acetone (5 × 30 mL). The gel was dried under reduced pressure (100 Pa) at 120 °C and finally ground into a fine powder for characterization.

3.2. Characterization Methods

Fourier transform infrared (FTIR) spectra were recorded in attenuated total reflection (ATR) mode using a Spectrum II spectrometer (PerkinElmer, ICGM, Montpellier, France).

Powder X-ray diffraction (XRD) patterns were recorded using a Bruker D8 diffractometer with CuKα radiation (λ = 0.15418 nm), using a Ni filter to suppress CuKβ radiation.

Scanning electron microscopy (SEM) images were acquired on a Hitachi S-4800 microscope and energy-dispersive X-ray spectroscopy (SEM–EDX) was conducted using an Oxford Instruments X-MaxN SDD detector (ICGM, Montpellier, France).

Solid-state ^{31}P magic angle spinning (MAS) NMR experiments were performed on a Varian VNMRS 400 MHz (9.4 T) spectrometer equipped with a 3.2 mm T3 HXY MAS probe. Single-pulse experiments were carried out with a spinning speed of 20 kHz, a 90° excitation pulse of 3 μs, a recycle delay of 30 s, and SPINAL-64 ^1H decoupling at 100 kHz. Two hundred transients were accumulated. The ^{31}P chemical shifts were referenced externally to hydroxyapatite ($Ca_{10}(PO_4)_6(OH)_2$) at 2.8 ppm relative to 85% H_3PO_4.

Solid-state ^{13}C cross-polarization magic angle spinning (CPMAS) NMR spectra were recorded on a Varian VNMRS 300 MHz spectrometer using a 3.2 mm double-resonance T3 probe, with a spinning speed of 12 kHz.

The textural properties of the materials were investigated by nitrogen adsorption–desorption measurements at −196 °C using a 3Flex surface analyzer (Micromeritics). Prior to analysis, the samples were degassed under vacuum (1.33×10^{-3} mbar) at 250 °C for 15 h. Specific surface areas (SSAs) were determined using the Brunauer–Emmett–Teller (BET) model within a relative pressure (P/P_0) range of 10^{-5} to 0.1. The total pore volume was estimated from the amount of nitrogen adsorbed at a relative pressure of 0.99. Microporous surface area and volume (for pores with widths w < 2 nm) were evaluated using the t-Plot method in the linear P/P_0 range of 0.2–0.5. Pore size distributions were determined by the Barrett–Joyner–Halenda (BJH) method, applied to the desorption branch for pores in the 2–50 nm range, and by the Density Functional Theory (DFT) method for pores in the 1–5 nm range.

The thermal stability of the materials was assessed by thermogravimetric analysis (TGA) under synthetic air using a STA 409 PC Luxx thermal analyzer (Netzsch, ICGM, Montpellier, France). Samples were placed in alumina crucibles and heated to 650 °C at a rate of 5 °C·min^{-1}.

4. Conclusions

In summary, we describe a simple, general, low-cost, and original synthesis of hybrid mesoporous materials via a template-free non-hydrolytic route. These hybrids contain TiO_2 nanodomains bridged by different bisphosphonates. All materials obtained have high specific surface areas and large pore diameters, with either flexible or rigid linkers. The unique combination of accessible functionality and outstanding aqueous stability makes these bridged mesoporous titania–bisphosphonates promising materials. This family of hybrids is complementary to existing organosilicas, MOFs, and metal phosphonates, depending on uses. This opens exciting opportunities in various advanced applications, such as heterogeneous catalysis, drug delivery, ion exchange and adsorption/separation. Their effectiveness in these applications will be investigated in due course.

Supplementary Materials: The following supporting information can be downloaded at: https://www.mdpi.com/article/10.3390/molecules30112459/s1, Figure S1: ^{13}C MAS NMR magic angle spinning solid-state NMR spectra of the bC_{12}P-Ti and bPy_2P-Ti hybrid materials; Figure S2: SEM images of bPh_2P-Ti; Figure S3: ATR-FTIR spectra of the BMOP hybrid materials; Figure S4: TGA curves of flexible (a) and rigid (b) BMOPs.

Author Contributions: All authors have contributed substantially to the work reported. Conceptualization and methodology, J.G.A.; Investigation, E.G., Z.S., Y.W. and K.B.; Formal analysis, E.G., Z.S., Y.W. and J.R.; Writing—Original draft preparation, J.G.A.; Supervision, J.G.A.; Funding acquisition and project administration, J.G.A. All authors have read and agreed to the published version of the manuscript.

Funding: This research was funded by the "Agence Nationale de la Recherche", grant number ANR-24-CE09-6614-01.

Institutional Review Board Statement: Not applicable.

Informed Consent Statement: Not applicable.

Data Availability Statement: The data presented in this study are available on request from the corresponding author.

Acknowledgments: The authors thank the Université de Montpellier and the Centre National de la Recherche Scientifique (CNRS) for financial support. The authors also express their gratitude to several internship students for their participation in the project: Lucas Rogier, Jarod Vincent, and Chloé Alexanko.

Conflicts of Interest: The authors declare no conflicts of interest.

References

1. Faustini, M.; Nicole, L.; Ruiz-Hitzky, E.; Sanchez, C. History of Organic–Inorganic Hybrid Materials: Prehistory, Art, Science, and Advanced Applications. *Adv. Funct. Mater.* **2018**, *28*, 1704158. [CrossRef]
2. Sanchez, C.; Julián, B.; Belleville, P.; Popall, M. Applications of Hybrid Organic–Inorganic Nanocomposites. *J. Mater. Chem.* **2005**, *15*, 3559–3592. [CrossRef]
3. Hoffmann, F.; Cornelius, M.; Morell, J.; Fröba, M. Silica-Based Mesoporous Organic–Inorganic Hybrid Materials. *Angew. Chem. Int. Ed.* **2006**, *45*, 3216–3251. [CrossRef]
4. Mizoshita, N.; Tani, T.; Inagaki, S. Syntheses, Properties and Applications of Periodic Mesoporous Organosilicas Prepared from Bridged Organosilane Precursors. *Chem. Soc. Rev.* **2011**, *40*, 789–800. [CrossRef] [PubMed]

5. Park, S.S.; Santha Moorthy, M.; Ha, C.-S. Periodic Mesoporous Organosilicas for Advanced Applications. *NPG Asia Mater.* **2014**, *6*, e96. [CrossRef]
6. Urata, C.; Yamada, H.; Wakabayashi, R.; Aoyama, Y.; Hirosawa, S.; Arai, S.; Takeoka, S.; Yamauchi, Y.; Kuroda, K. Aqueous Colloidal Mesoporous Nanoparticles with Ethenylene-Bridged Silsesquioxane Frameworks. *J. Am. Chem. Soc.* **2011**, *133*, 8102–8105. [CrossRef]
7. Horike, S.; Shimomura, S.; Kitagawa, S. Soft Porous Crystals. *Nat. Chem.* **2009**, *1*, 695–704. [CrossRef]
8. Furukawa, H.; Cordova, K.E.; O'Keeffe, M.; Yaghi, O.M. The Chemistry and Applications of Metal-Organic Frameworks. *Science* **2013**, *341*, 1230444. [CrossRef]
9. Machac, P.; Styskalik, A.; Moravec, Z.; Pinkas, J. Non-Hydrolytic Sol-Gel Synthesis of Zirconium Phosphonates with Controlled Mesoporosity. *Microporous Mesoporous Mater.* **2023**, *362*, 112787. [CrossRef]
10. Popov, K.; Oshchepkov, M.; Tkachenko, S.; Sergienko, V.; Oshchepkov, A. Bisphosphonates: Synthesis, Structures, Properties, Medical and Industrial Applications. *J. Mol. Liq.* **2022**, *351*, 118619. [CrossRef]
11. Pawlak, B.; Marchal, W.; Ramesha, B.M.; Joos, B.; Calvi, L.; D'Haen, J.; Ruttens, B.; Hardy, A.; Meynen, V.; Adriaensens, P.; et al. Hybrid Porous Titania Phosphonate Networks with Different Bridging Functionalities: Synthesis, Characterization, and Evaluation as Efficient Solvent Separation Materials. *Microporous Mesoporous Mater.* **2022**, *341*, 112080. [CrossRef]
12. Machac, P.; Alauzun, J.; Styskalik, A.; Debecker, D.; Mutin, P.H.; Pinkas, J. Synthesis of High Surface Area Aluminophosphate and -Phosphonate Xerogels by Non-Hydrolytic Sol-Gel Reactions. *Microporous Mesoporous Mater.* **2021**, *311*, 110682. [CrossRef]
13. Bao, S.-S.; Qin, M.-F.; Zheng, L.-M. Metal Phosphonates Incorporating Metalloligands: Assembly, Structures and Properties. *Chem. Commun.* **2020**, *56*, 12090–12108. [CrossRef] [PubMed]
14. Wang, Q.; Astruc, D. State of the Art and Prospects in Metal–Organic Framework (MOF)-Based and MOF-Derived Nanocatalysis. *Chem. Rev.* **2020**, *120*, 1438–1511. [CrossRef]
15. Lim, D.-W.; Kitagawa, H. Proton Transport in Metal-Organic Frameworks. *Chem. Rev.* **2020**, *120*, 8416–8467. [CrossRef]
16. Howarth, A.J.; Liu, Y.; Li, P.; Li, Z.; Wang, T.C.; Hupp, J.T.; Farha, O.K. Chemical, Thermal and Mechanical Stabilities of Metal–Organic Frameworks. *Nat. Rev. Mater.* **2016**, *1*, 15018. [CrossRef]
17. Schmitt Pauly, C.; Genix, A.-C.; Alauzun, J.G.; Guerrero, G.; Appavou, M.-S.; Pérez, J.; Oberdisse, J.; Mutin, P.H. Simultaneous Phase Transfer and Surface Modification of TiO_2 Nanoparticles Using Alkylphosphonic Acids: Optimization and Structure of the Organosols. *Langmuir* **2015**, *31*, 10966–10974. [CrossRef]
18. Mutin, P.H.; Guerrero, G.; Vioux, A. Hybrid Materials from Organophosphorus Coupling Molecules. *J. Mater. Chem.* **2005**, *15*, 3761–3768. [CrossRef]
19. Zhu, Y.-P.; Ma, T.-Y.; Liu, Y.-L.; Ren, T.-Z.; Yuan, Z.-Y. Metal Phosphonate Hybrid Materials: From Densely Layered to Hierarchically Nanoporous Structures. *Inorg. Chem. Front.* **2014**, *1*, 360–383. [CrossRef]
20. Gagnon, K.J.; Perry, H.P.; Clearfield, A. Conventional and Unconventional Metal–Organic Frameworks Based on Phosphonate Ligands: MOFs and UMOFs. *Chem. Rev.* **2012**, *112*, 1034–1054. [CrossRef]
21. Zhu, Y.-P.; Ren, T.-Z.; Yuan, Z.-Y. Insights into Mesoporous Metal Phosphonate Hybrid Materials for Catalysis. *Catal. Sci. Technol.* **2015**, *5*, 4258–4279. [CrossRef]
22. Luca, V.; Tejada, J.J.; Vega, D.; Arrachart, G.; Rey, C. Zirconium(IV)-Benzene Phosphonate Coordination Polymers: Lanthanide and Actinide Extraction and Thermal Properties. *Inorg. Chem.* **2016**, *55*, 7928–7943. [CrossRef] [PubMed]
23. Styskalik, A.; Skoda, D.; Moravec, Z.; Babiak, M.; Barnes, C.E.; Pinkas, J. Control of Micro/Mesoporosity in Non-Hydrolytic Hybrid Silicophosphate Xerogels. *J. Mater. Chem. A* **2015**, *3*, 7477–7487. [CrossRef]
24. Vasylyev, M.; Neumann, R. Preparation, Characterizaton, and Catalytic Aerobic Oxidation by a Vanadium Phosphonate Mesoporous Material Constructed from a Dendritic Tetraphosphonate. *Chem. Mater.* **2006**, *18*, 2781–2783. [CrossRef]
25. Kimura, T. A New Family of Nonsiliceous Porous Hybrids from Bisphosphonates. *J. Nanosci. Nanotechnol.* **2013**, *13*, 2461–2470. [CrossRef]
26. Ma, T.-Y.; Li, H.; Tang, A.-N.; Yuan, Z.-Y. Ordered, Mesoporous Metal Phosphonate Materials with Microporous Crystalline Walls for Selective Separation Techniques. *Small* **2011**, *7*, 1827–1837. [CrossRef]
27. Veliscek-Carolan, J.; Rawal, A.; Luca, V.; Hanley, T.L. Zirconium Phosphonate Sorbents with Tunable Structure and Function. *Microporous Mesoporous Mater.* **2017**, *252*, 90–104. [CrossRef]
28. Li, H.; Sun, Y.; Yuan, Z.-Y.; Zhu, Y.-P.; Ma, T.-Y. Titanium Phosphonate Based Metal–Organic Frameworks with Hierarchical Porosity for Enhanced Photocatalytic Hydrogen Evolution. *Angew. Chem. Int. Ed.* **2018**, *57*, 3222–3227. [CrossRef]
29. Dines, M.B.; Cooksey, R.E.; Griffith, P.C.; Lane, R.H. Mixed-Component Layered Tetravalent Metal Phosphonates/Phosphates as Precursors for Microporous Materials. *Inorg. Chem.* **1983**, *22*, 1003–1004. [CrossRef]
30. Wang, Y.; Alauzun, J.G.; Mutin, P.H. Water-Stable, Nonsiliceous Hybrid Materials with Tunable Porosity and Functionality: Bridged Titania-Bisphosphonates. *Chem. Mater.* **2020**, *32*, 2910–2918. [CrossRef]

31. Van Velthoven, N.; Wang, Y.; Van Hees, H.; Henrion, M.; Bugaev, A.L.; Gracy, G.; Amro, K.; Soldatov, A.V.; Alauzun, J.G.; Mutin, P.H.; et al. Heterogeneous Single-Site Catalysts for C–H Activation Reactions: Pd(II)-Loaded S,O-Functionalized Metal Oxide-Bisphosphonates. *ACS Appl. Mater. Interfaces* **2020**, *12*, 47457–47466. [CrossRef] [PubMed]
32. Thommes, M.; Kaneko, K.; Neimark, A.V.; Olivier, J.P.; Rodriguez-Reinoso, F.; Rouquerol, J.; Sing, K.S.W. Physisorption of Gases, with Special Reference to the Evaluation of Surface Area and Pore Size Distribution (IUPAC Technical Report). *Pure Appl. Chem.* **2015**, *87*, 1051–1069. [CrossRef]
33. Groen, J.C.; Peffer, L.A.A.; Pérez-Ramírez, J. Pore Size Determination in Modified Micro- and Mesoporous Materials. Pitfalls and Limitations in Gas Adsorption Data Analysis. *Microporous Mesoporous Mater.* **2003**, *60*, 1–17. [CrossRef]
34. Brunauer, S.; Emmett, P.H.; Teller, E. Adsorption of Gases in Multimolecular Layers. *J. Am. Chem. Soc.* **1938**, *60*, 309–319. [CrossRef]
35. Wang, Y.; Mutin, P.H.; Alauzun, J.G. One-Step Nonhydrolytic Sol–Gel Synthesis of Mesoporous TiO_2 Phosphonate Hybrid Materials. *Beilstein J. Nanotechnol.* **2019**, *10*, 356–362. [CrossRef]
36. Wang, Y.; Kim, S.; Louvain, N.; Alauzun, J.G.; Mutin, P.H. Acetic Anhydride as an Oxygen Donor in the Non-Hydrolytic Sol-Gel Synthesis of Mesoporous TiO_2 with High Electrochemical Lithium Storage Performances. *Chem. Weinh. Bergstr. Ger.* **2019**, *25*, 4767–4774. [CrossRef]
37. Guerrero, G.; Mutin, P.H.; Vioux, A. Anchoring of Phosphonate and Phosphinate Coupling Molecules on Titania Particles. *Chem. Mater.* **2001**, *13*, 4367–4373. [CrossRef]
38. Guerrero, G.; Mutin, P.H.; Vioux, A. Mixed Nonhydrolytic/Hydrolytic Sol−Gel Routes to Novel Metal Oxide/Phosphonate Hybrids. *Chem. Mater.* **2000**, *12*, 1268–1272. [CrossRef]

Disclaimer/Publisher's Note: The statements, opinions and data contained in all publications are solely those of the individual author(s) and contributor(s) and not of MDPI and/or the editor(s). MDPI and/or the editor(s) disclaim responsibility for any injury to people or property resulting from any ideas, methods, instructions or products referred to in the content.

molecules

MDPI

Article

Mechanical and Thermal Performance of In-Situ Synthesized PDMS-SiO_2 Composite as Electrical Insulating Coatings

Aldo Cordoba [1], Rossana Faride Vargas-Coronado [2], Rodrigo Velázquez-Castillo [1], Juan Valerio Cauich-Rodríguez [2,*] and Karen Esquivel [1,*]

1 Graduate and Research Division, Engineering Faculty, Universidad Autónoma de Querétaro, Cerro de las Campanas, Santiago de Querétaro 76010, Querétaro, Mexico; acordoba07@alumnos.uaq.mx (A.C.); rodrigo.velazquez@uaq.mx (R.V.-C.)

2 Centro de Investigación Científica de Yucatán, Unidad de Materiales, C. 43 No. 130 x 32 y 34, Col. Chuburná de Hidalgo, Mérida 97205, Yucatán, Mexico; ross@cicy.mx

* Correspondence: jvcr@cicy.mx (J.V.C.-R.); karen.esquivel@uaq.mx (K.E.); Tel.: +52-442-192-1200 (ext. 65401) (K.E.)

Abstract: Polydimethylsiloxane (PDMS) has been extensively employed in electrical insulation applications owing to its excellent thermal stability, hydrophobicity, and dielectric properties. However, its inherent mechanical limitations require structural reinforcement to enhance its performance under more demanding operational conditions. In this study, the mechanical, thermal, and surface properties of PDMS-SiO_2 nanocomposites synthesized via in situ sol–gel process was systematically investigated. The influence of different SiO_2 nanoparticle concentrations (5, 10, and 15 wt%), sol–gel catalyst type (acidic and alkaline), and tetraethyl orthosilicate (TEOS) crosslinking agent ratios (15:1, 10:1, 5:1) was evaluated. Tensile mechanical testing, dynamic mechanical analysis (DMA), and thermogravimetric analysis (TGA) revealed that the incorporation of SiO_2 notably improved both the mechanical strength and thermal stability of the composites. The 5-15b and 10-15a composites exhibited the highest tensile stress and viscoelastic modulus among all samples. Furthermore, the composites retained key functional properties, including hydrophobicity, high volumetric electrical resistivity (~10^{11} $\Omega \cdot$cm), and strong adhesion. These findings confirm the potential of in situ PDMS-SiO_2 nanocomposites for use as high-performance insulating coatings in advanced electrical applications.

Keywords: in situ sol–gel; hybrid O/I systems; wettability behavior; thermal behavior; mechanical properties

1. Introduction

Polydimethylsiloxane (PDMS) is a widely used polymer in various applications due to its remarkable flexibility, chemical stability, hydrophobicity, and thermal resistance [1]. Its high electrical insulation and chemical stability against environmental factors such as radiation and humidity make it an ideal candidate for protective coatings in electronic and electrical applications, including high-voltage cables and insulators in power plants [2]. However, PDMS exhibits limitations such as low mechanical strength and hardness, which requires its integration with additional materials to enhance its durability under external conditions and extend its operational life [3].

Several reinforcement strategies have been proposed to enhance the mechanical performance of PDMS when applied as an insulating coating. Among the most extensively studied are nanostructures based on metal oxides such as titanium dioxide (TiO_2) [4,5], alumina (Al_2O_3) [6], and silicon dioxide (SiO_2) [7,8], as well as carbon-based nanomaterials including carbon nanotubes (CNTs) [9] and graphene [10]. Previous investigations have demonstrated that TiO_2 nanoparticles can enhance the elastic modulus and tensile strength of PDMS matrices; however, significant agglomeration tendencies and reduced flexibility have been reported, limiting their applicability [11]. Similarly, CNT-reinforced PDMS composites exhibit substantial improvements in mechanical strength and electrical conductivity; nevertheless, dispersion difficulties and the tendency to create localized stress concentrations negatively impact its tensile strength [12].

Nanocomposite materials based on PDMS and SiO_2 have attracted considerable attention due to their improved mechanical and thermal properties [4,13]. The incorporation of SiO_2 into the PDMS matrix did not alter the chemical environment or the spatial arrangement of the polymeric network, owing to the chemical similarity between the matrix and the reinforcement [7]. This compatibility enabled the preservation of the elastomeric behavior of PDMS and facilitated an effective dispersion of the reinforcing phase within the composite structure.

Despite the multiple nanostructures that have been shown to modify and improve the mechanical properties of these polymers, traditional approaches for the fabrication of PDMS composites have primarily relied on ex situ synthesis (pre-formed nanoparticles were dispersed into the polymer matrix) which ends up limiting the addition of reinforcers due to the difficulty of dispersion and agglomeration in high-viscosity polymers [14]. That is why the use of high-energy mixing techniques to achieve adequate dispersion is required [15], thereby limiting the maximum concentration of reinforcements that could be effectively incorporated. These conditions have ultimately led to suboptimal interfacial interactions between the polymer matrix and the nanoparticles, adversely affecting the final performance of the composite materials [9].

To address these limitations, recent research efforts have been focused on the development of in situ synthesis techniques, which allowed the formation of nanoparticles directly within the polymeric matrix [14]. In situ synthesis methods include emulsion polymerization [16], hydrothermal-assisted synthesis [17], and the sol–gel process [7]. Among these, the sol–gel method has offered several advantages, including enhanced nanoparticle dispersion, adaptability to diverse polymer matrix synthesis conditions, and tunability of the physicochemical properties of the reinforcements by modifying key synthesis parameters such as catalyst type, pH, and precursor concentration [18]. For SiO_2 nanoparticles, the influence of sol–gel catalysts was found to be a critical factor in determining the size, morphology, and surface chemistry of the synthesized nanoparticles, thereby potentially impacting the overall performance of the resulting PDMS composites [19].

Although significant advances have been achieved in the development of PDMS–SiO_2 composites [7], comprehensive investigations addressing the influence of key synthesis variables—such as nanoparticle concentration, sol–gel catalyst selection, and crosslinking degree—on the mechanical and thermal performance of these materials remain limited. Previous studies have indicated that increasing nanoparticle content can improve the mechanical reinforcement of PDMS; however, excessive nanoparticle loading has been associated with agglomeration phenomena, which may detrimentally affect the uniformity and performance of the composite material [4,20]. Furthermore, the effect of sol–gel catalyst conditions on the structural homogeneity and interfacial interactions within the PDMS matrix has not yet been fully elucidated and continues to be the subject of ongoing research efforts.

Previous studies have emphasized the importance of combining crosslinking processes and nanoparticle incorporation; however, the competitive nature of the nanoparticle synthesis and polymer crosslinking reactions often limited the extent of polymer–nanoparticle interaction and, consequently, the final properties of the composites [21]. Additionally, although higher degrees of crosslinking generally enhanced the mechanical strength and thermal stability of PDMS-based systems, they also tended to induce excessive rigidity and reduce flexibility, which could negatively affect the performance of the material when applied as a coating [22].

In addition to the enhancement of mechanical properties, attributes such as hydrophobicity, electrical insulation, and adhesion are crucial for the optimal performance of these composites when employed as insulating coatings [1,3].

Hydrophobicity and superficial adhesion play a critical role in the durability and protective capability of insulating coatings, as they help prevent moisture absorption, contamination, and delamination which degrade electrical insulation [23]. Similarly, maintaining electrical insulation properties is essential to ensure the suitability of these composites for electrical applications [24]. The formation of composites through the incorporation of SiO_2 reinforcements and variations in crosslinking conditions may impact these properties and warrants further investigation.

This study continues the evaluation of the in situ synthesis approach for PDMS–SiO_2 composites previously developed by the research group. In earlier work, the physicochemical characterization of sol–gel-synthesized SiO_2 nanoparticles and their corresponding composites was conducted through FTIR, Raman, XRD, and XPS analyses. The present investigation expands upon these findings by examining the influence of synthesis parameters—specifically nanoparticle concentration, sol–gel catalyst type (acidic and alkaline), and crosslinking agent (TEOS) ratio—on the mechanical and thermal properties of the resulting composites. The objective is to gain a deeper understanding of how these variables modify the internal chain mobility mechanisms within the polymer matrix. The results contribute to the advancement of polymer nanocomposites for electrical insulation applications and may serve as a foundation for future developments in nanoreinforced functional coating materials.

This study continues the evaluation of the in situ synthesis approach for PDMS–SiO_2 composites previously developed by the research group. In earlier work, the physicochemical characterization of sol–gel-synthesized SiO_2 nanoparticles and their corresponding composites was conducted through FTIR, Raman, XRD, and XPS analyses [7]. The present investigation expands upon these findings by examining the influence of synthesis parameters (nanoparticle concentration, sol–gel catalyst type (acidic and alkaline), and crosslinking agent ratio) on the mechanical and thermal properties of the resulting composites. The objective is to gain a deeper understanding of how these variables modify the internal chain mobility mechanisms within the polymer matrix. The results contribute to the advancement of polymer nanocomposites for electrical insulation applications and may serve as a foundation for future developments in nanoreinforced functional coating materials.

2. Results

To guarantee the accurate identification and analysis of the different study groups, a specific nomenclature was established. Each composite was designated by first indicating the PDMS:TEOS weight ratio (15, 10, or 5), followed by the SiO_2 reinforcement content (0, 5, 10, or 15 wt%), and finally by the synthesis medium, with "a" denoting acidic conditions and "b" representing alkaline conditions. For instance, the composite labeled 5-15b corresponds to a PDMS:TEOS weight ratio of 5:1, reinforced with 15 wt% of SiO_2 nanoparticles synthesized under alkaline conditions.

2.1. PDMS-%SiO_2 Mechanical and Thermal Properties

2.1.1. Tensile Mechanical Analysis

Tensile tests were conducted on the various PDMS-%SiO_2 polymeric composites to determine the changes in mechanical behavior caused by different concentrations of SiO_2 reinforcer (0%, 5%, 10%, and 15%) obtained via sol–gel synthesis in acidic (a) and alkaline (b) media. In Figure 1, the stress–strain profiles for the different study groups are presented, revealing a notable increase in both ultimate tensile stress and strain compared to the control unreinforced polymers (15-0, 10-0, and 5-0). In the unreinforced polymers, two primary processes were observed—one occurring between 0 and 0.4 MPa and another above 0.4 MPa—which corresponded to the elastic and plastic deformation stages, respectively [1].

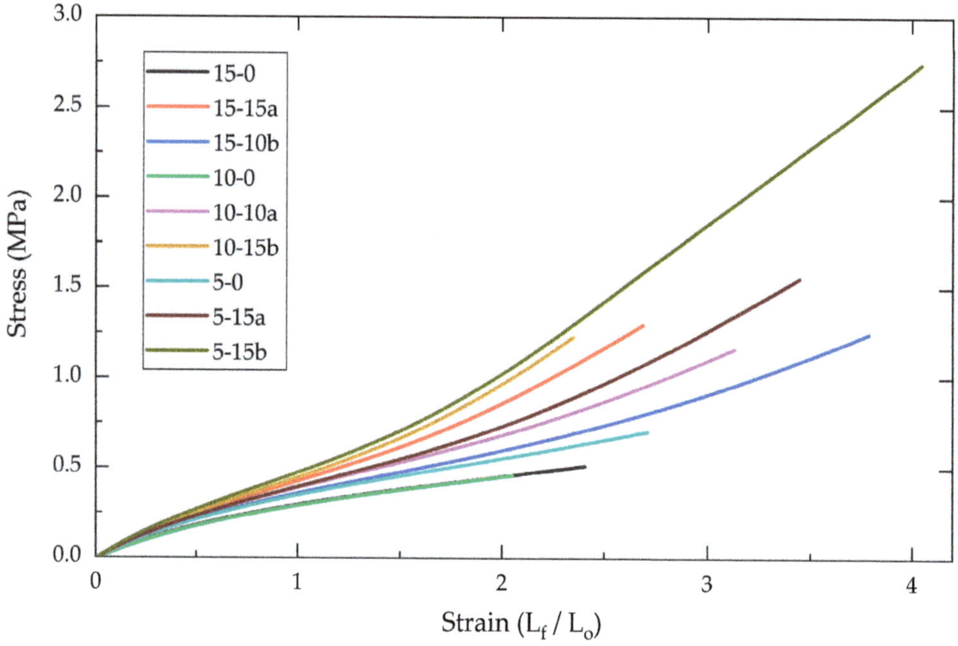

Figure 1. Stress–strain profiles of PDMS-%SiO_2 composites for SiO_2-a and SiO_2-b reinforcers.

In the reinforced composites (PDMS-%SiO_2), a third creep process was observed at 1.5 MPa, which is attributed to a plastic deformation mechanism commonly encountered in reinforced polymers and related to matrix–particle interactions [7,11].

Based on the stress–strain profiles, the elastic modulus, ultimate strain, and fracture stress for each reinforced composite were calculated and presented in Table 1. In Figure 2, a graphical representation of the progressive changes in these mechanical properties (y-axis) is provided as the reinforcement concentration increased from 0% to 15% (x-axis).

For the unreinforced materials (15-0, 10-0, and 5-0), the elastic modulus increased with the PDMS:TEOS weight ratio, ranging from 0.393 to 0.496 MPa, corresponding to a 26% increase for the control group with the highest crosslinking concentration (5-0) [25]. Among the reinforced composite groups, the elastic modulus exhibited a positive correlation with the concentration of SiO_2 reinforcement. The composites exhibiting the highest elastic modulus values were 5-15b and 10-15b, with values of 0.60 ± 0.02 and 0.59 ± 0.01 MPa, respectively. These values represented increases of 20% and 51% relative to its unreinforced control groups (5-0 and 10-0).

In contrast to the elastic modulus, the ultimate strain of the composites did not exhibit a discernible relationship with the SiO_2 concentration. Within the unreinforced polymer samples, the 10-0 group exhibited the lowest strain at fracture, with a dimension ratio (D_f/D_0) of 2.12 (corresponding to a deformation of 212%). Among the reinforced composites, the highest dimension ratios were observed in 5-15b and 10-15a groups, with ratios of 4.08 and 3.88, respectively, representing increases of 42% and 83% compared to the unreinforced polymers.

The ultimate tensile stress property exhibited a rising trend as the concentration of SiO_2 reinforcements increased. The unreinforced PDMS composites demonstrated fracture stress values ranging from 0.471 to 0.731 MPa, with the 10-0 group displaying the lowest fracture resistance.

For composites with low SiO_2 concentrations (5a and 5b), fracture stress values ranged from 0.484 to 0.901 MPa. Gradually, the fracture stress increased from 0.934 to 1.281 MPa for composites with a 10% reinforcement concentration (10a and 10b, respectively). At the highest reinforcement concentration (15% SiO_2), most composites exhibited a considerable increase in ultimate tensile stress, thereby differing from the behavior at lower concentrations [26]. The composites with the highest fracture stress for each reinforcer (SiO_2-a and SiO_2-b) were 10-15a and 5-15b (1.94 and 2.71 MPa, respectively). This value represented an increase of 371% and 412% in comparison with its control groups.

Table 1. Tensile mechanical test properties for PDMS-%SiO_2 composites.

			Elastic Modulus (MPa)			
PDMS:TEOS	15		10		5	
SiO_2 [1] (%w)	a	b	a	b	a	b
0	0.39 ± 0.01	0.39 ± 0.01	0.39 ± 0.01	0.39 ± 0.01	0.50 ± 0.013	0.50 ± 0.01
5	0.46 ± 0.02	0.47 ± 0.004	0.51 ± 0.03	0.48 ± 0.02	0.49 ± 0.010	0.48 ± 0.01
10	0.50 ± 0.001	0.50 ± 0.01	0.52 ± 0.02	0.51 ± 0.01	0.50 ± 0.009	0.52 ± 0.01
15	0.56 ± 0.01	0.46 ± 0.01	0.54 ± 0.03	0.59 ± 0.01	0.52 ± 0.023	0.60 ± 0.02
			Ultimate Strain (MPa)			
PDMS:TEOS	15		10		5	
SiO_2 [1] (%w)	a	b	a	b	a	b
0	3.00 ± 0.15	3.00 ± 0.15	2.12 ± 0.26	2.12 ± 0.26	2.87 ± 0.54	2.87 ± 0.54
5	1.70 ± 0.35	3.11 ± 0.45	1.88 ± 0.32	3.25 ± 0.46	3.00 ± 0.18	3.48 ± 0.45
10	2.91 ± 0.24	3.80 ± 0.31	3.16 ± 0.13	3.48 ± 0.38	3.52 ± 0.32	2.76 ± 0.19
15	2.79 ± 0.28	2.91 ± 0.05	3.88 ± 0.56	2.46 ± 0.33	3.40 ± 0.35	4.08 ± 0.19
SiO_2 [1] (%w)			**Ultimate Tensile Stress (MPa)**			
PDMS:TEOS	15		10		5	
[C] SiO_2 [1]	a	b	a	b	a	b
0	0.59 ± 0.07	0.59 ± 0.07	0.47 ± 0.04	0.47 ± 0.04	0.73 ± 0.02	0.73 ± 0.03
5	0.48 ± 0.04	0.75 ± 0.09	0.57 ± 0.05	0.85 ± 0.14	0.82 ± 0.09	0.90 ± 0.11
10	0.93 ± 0.11	1.24 ± 0.13	1.15 ± 0.03	1.28 ± 0.23	1.28 ± 0.10	1.00 ± 0.11
15	1.34 ± 0.22	0.76 ± 0.05	1.94 ± 0.22	1.34 ± 0.27	1.53 ± 0.30	2.71 ± 0.16

[1] Rows correspond to the SiO_2 concentration, and columns correspond to the catalyst type: acidic (a) or alkaline (b).

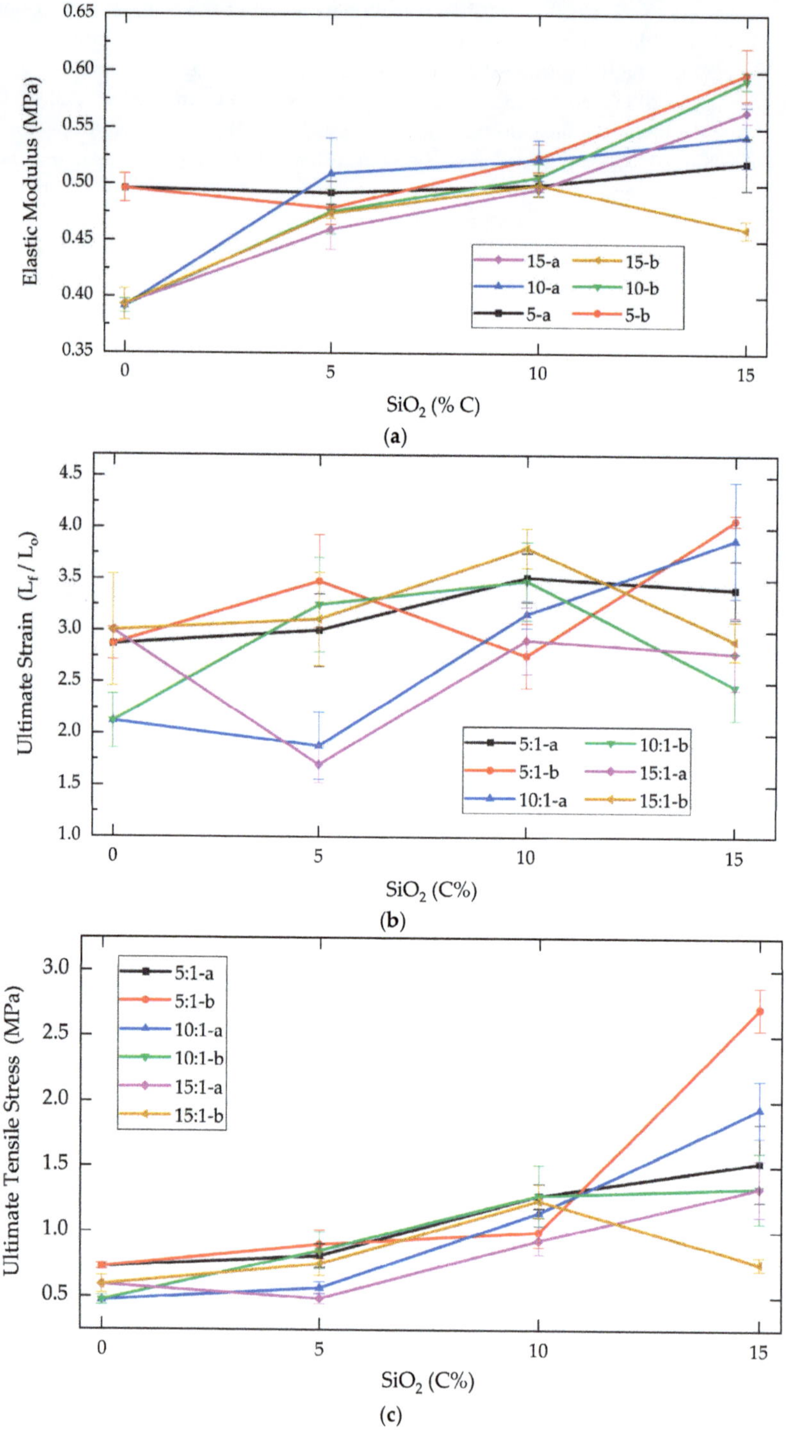

Figure 2. Mechanical properties of PDMS-%SiO_2 composites: (**a**) elastic modulus, (**b**) ultimate deformation, and (**c**) ultimate tensile stress.

2.1.2. Dynamic Mechanic Analysis

In addition to the tensile tests, dynamic mechanical analyses were conducted to elucidate the mechanical behavior of the various composites by examining variations in their viscoelastic properties and thermal transitions over a temperature range [27]. DMA tests were performed on the composites within a temperature window from−130 to 80 °C. Table 2 presents the thermal transition processes, including the glass transition temperature (T_g), as well as the storage modulus (G′), loss modulus (G″), and tan δ (G′/G″) measured at 30 °C for each study group.

Table 2. Dynamic mechanical analysis test results for PDMS-%SiO_2 composites.

	G″ (MPa)	G′ (MPa)	Tan δ	Tg (°C) [1]
15-0	0.035	0.970	0.036	−66.79
15-5a	0.013	1.508	0.008	−70.14
15-10a	0.058	1.080	0.053	−60.72
15-15a	0.086	1.521	0.057	−59.74
15-5b	0.062	1.014	0.061	−59.81
15-10b	0.063	1.160	0.054	−59.88
15-15b	0.014	1.267	0.011	−72.72
10-0	0.064	1.097	0.059	−69.31
10-5a	0.064	1.191	0.053	−79.25
10-10a	0.063	1.103	0.057	−66.60
10-15a	0.062	1.161	0.054	−67.00
10-5b	0.064	1.059	0.060	−61.36
10-10b	0.071	1.101	0.057	−61.15
10-15b	0.071	1.301	0.053	−61.05
5-0	0.058	0.972	0.060	−61.77
5-5a	0.060	1.035	0.058	−65.21
5-10a	0.068	1.081	0.063	−64.31
5-15a	0.076	1.174	0.064	−60.27
5-5b	0.060	1.246	0.048	−63.34
5-10b	0.097	1.267	0.077	−62.34
5-15b	0.077	1.398	0.055	−52.54

[1] T_g was calculated by extrapolating the first mechanical transformation process from the G′ vs. T.

For all composite groups, a viscoelastic behavior characteristic of the PDMS polymer matrix was observed, with a larger contribution from the solid elastic response (G′) behavior after the crosslinking process [28]. A gradual increase in the storage modulus was observed within the incorporation of SiO_2 nanoparticles. The composites that exhibit the highest reinforcement concentrations were 15-15a, 10-15b, and 5-15b.

Glass transition temperature of composites varied as a function of the crosslinking degree and reinforcement concentration, presenting the composites 15-15a, 10-15b, and 5-15b the highest transition temperatures. Composite 5-15b exhibited the most pronounced temperature increase (−52.54 °C), representing an increase of over 9 °C relative to the control polymer 5-0 (−61.77 °C). This behavior can be attributed to changes in the crosslinking degree, reducing the mobility of the polymer chains [1].

2.1.3. Thermogravimetric Analysis

Once the mechanical tests (static and dynamic) had been performed, it was determined that the composites exhibiting the highest mechanical strength for each type of SiO_2 reinforcement were 10-15a and 5-15b. Analyses of thermal properties of these composites were performed for a more comprehensive understanding of the interaction between the polymeric system and the selected variables (sol–gel catalyst, SiO_2 concentration, and crosslinking agent concentration) on the thermal mobility of the polymer chains and thermal stability [29]. To identify these progressive changes in the thermal behavior of the composites with superior mechanical performance (10-15a and 5-15b), thermogravimetric analyses were performed.

Figure 3a shows the weight loss profiles of the best mechanical composite subgroups, from the control groups 10-0 and 5-0 up to the highest concentrations, 10-15a and 5-15b. In Figure 3b, the weight loss rate of the composites, as the temperature increased to 700 °C, is represented.

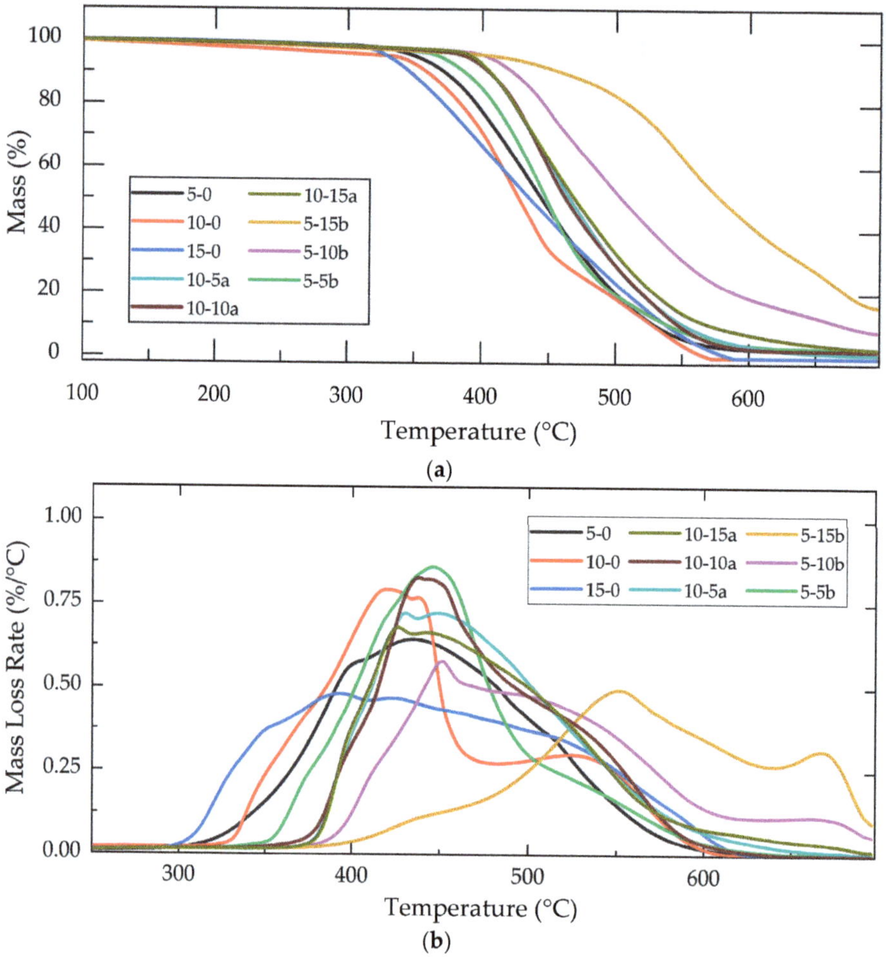

Figure 3. TGA thermograms of PDMS-%SiO_2 composites: (**a**) weight loss percentage and (**b**) derivative TGA (%/°C).

For direct comparison among the study groups, the onset of the primary degradation process was determined via linear fitting. The results for the different composites are presented in Table 3. Additionally, the weight loss rate, and percentage of residual mass after 700 °C were recorded.

Table 3. Thermogravimetric analysis for PDMS-%SiO_2 composites.

Composite	Intersection (100 °C)	Max. Mass Loss		Reisdual Mass 700 °C (%)
		T (°C)	Rate (%/°C)	
15-0	327	392	0.48	0
10-0	351	419	0.79	0
10-5a	391	447	0.72	1.17
10-10a	395	437	0.83	2.26
10-15a	393	425	0.68	2.94
5-0	358	435	0.65	2.27
5-5b	377	445	0.86	2.85
5-10b	405	451	0.58	8.35
5-15b	447	550	0.5	16.16

From mass loss rate values shown in Figure 2b, two principal degradation processes were identified across the temperature window of 300–700 °C for all groups. These degradation processes were attributed to the rupture of the crosslinking chains and the subsequent degradation of the primary PDMS chains [11].

An increase in the onset degradation temperature was observed in the control groups (15-0, 10-0, and 5-0) as the crosslinker concentration change; as the amount of TEOS increased, the degradation onset temperature rose from 327 to 358 °C. This behavior was attributed to the higher crosslinking degree; a more extensively crosslinked matrix requires greater energy for polymer chain mobility, resulting in higher degradation temperatures [30].

For the reinforced composites, variations in thermal behavior were observed upon employing different SiO_2 concentrations (0, 5, 10, and 15%) and sol–gel catalysts (SiO_2-a or SiO_2-b). Composites reinforced with SiO_2-a exhibited a slight increase in the onset degradation temperature relative to the unreinforced polymer, reaching up to 395 °C (a 12% increase relative to 10-0), without an apparent dependence on reinforcement concentration. Conversely, composites containing SiO_2-b particles demonstrated a gradual increase in the onset degradation temperature as the concentration of SiO_2-b reinforcement increased, rising from 377 °C (5-5b) to 447 °C (5-15b), representing an increase of up to 25% compared to the control polymer (5-0).

In Table 3, the percentage of residual weight following heating to 700 °C was also presented. For 10-b and 5-a groups, a gradual increase in the composite's residual weight was observed as the reinforcement concentration increased [10]. For 10-0 composition, zero residual mass was observed at 700 C; however, the polymer's residual mass gradually increased with SiO_2-a concentration, reaching 2.94 w% for 10-15a. Similarly, in the 5-0, a residual weight of 2.27% was recorded, which gradually increased with rising SiO_2-b concentration, reaching 16.16% for the composite with the highest reinforcement concentration (5-15b).

The higher residual mass observed in composites synthesized under alkaline conditions (SiO_2-b) compared to those synthesized under acidic conditions (SiO_2-a) can be attributed to the enhanced crosslinking degree promoted in these systems. As previously reported, alkaline sol–gel synthesis routes favor the retention of higher concentrations of unreacted TEOS and other precursors after sol–gel particle formation [7], which subsequently can participate in additional crosslinking reactions within the PDMS matrix. This increase in crosslinking degree contributes to improved thermal stability and superior mechanical properties, resulting from the synergistic effect of increased crosslinking and the incorporation of reinforcing nanoparticles.

2.2. PDMS-%SiO_2 Coating Characterization

Crosslinking efficiency, hydrophobicity, conductivity, and adhesion tests were conducted on composite subgroups selected from the mechanical and thermal analyses (5-15b and 10-15a). These evaluations were performed to assess the impact of the selected variables on the essential properties required for the application of these materials as insulating coatings [24].

2.2.1. Crosslinking Efficiency

Crosslinking efficiency of selected composites was estimated via gravimetric weight loss tests, determining if high reinforcement concentrations significantly altered the crosslinking yield of PDMS. Triplicate composite samples were collected and weighed in their dry state (P_i), following immersion in ethanol (P_h), and after the drying process (P_S). Using Equations (1)–(3), the solvent absorption percentage (A%), weight loss percentage of the coating due to the solvent (L_W), and the crosslinking efficiency (X_y) were calculated for the composites after removal of the remaining subproducts and unpolymerized chains. The results are presented in Table 4.

$$A = \frac{P_h - P_i}{P_s} \tag{1}$$

$$L_W = \frac{P_i - P_s}{P_i} \tag{2}$$

$$X_y = \frac{P_s}{P_i} \tag{3}$$

Table 4. Gravimetric test results for crosslinking efficiency on PDMS-SiO_2 selected composites.

Group	A (%)	L_W (%)	X_y (%)
15-0	3.65 ± 0.55	3.59 ± 1.04	96.41 ± 1.04
10-0	2.15 ± 0.81	1.71 ± 0.86	98.29 ± 0.86
5-0	2.23 ± 0.90	1.78 ± 0.97	98.22 ± 0.97
10-15a	1.12 ± 0.80	3.49 ± 0.34	96.51 ± 0.34
5-15b	0.98 ± 0.75	3.28 ± 0.64	96.72 ±0.64

Solvent absorption (*A*) in the various control composites exhibited a dependence on the PDMS:TEOS ratio, progressively decreasing from 3.65% to 2.23% as the crosslinker concentration increased. This reduced solvent absorption can be attributed to the concentration of the elastomer's crosslinking agent [30]. In composites 10-15a and 5-15b, a clear reduction in solvent absorption was observed relative to the control polymers, with values reaching less than half those of the unreinforced samples, with 1.12% and 0.98%, respectively.

The crosslinking efficiency (X_y), complementary to L_W, exhibited similar trends across the various composites, with all study groups falling within a range of 96.4–98.3%. Nonetheless, in the control groups, a higher concentration of TEOS crosslinker was associated with a slight increase in crosslinking efficiency, attributed to the presence of un-crosslinked polymer chains that were displaced by dissolvent [31]. Conversely, the reinforced groups exhibited less efficiency than those of the control groups, a trend attributed to a higher concentration of by-products generated during hydrolysis and condensation reactions, as previously demonstrated [7]. These by-products were displaced and eliminated from composites during the solvent absorption and drying processes.

2.2.2. Hydrophobicity and Electric Insulation

Hydrophobicity of composites was assessed by measuring the water contact angle in air on the surfaces of the selected PDMS composites (10-15a and 5-15b) as well as on control polymers. Images obtained from the contact angle tests are presented in Figure 4, and the measured angles in Table 5.

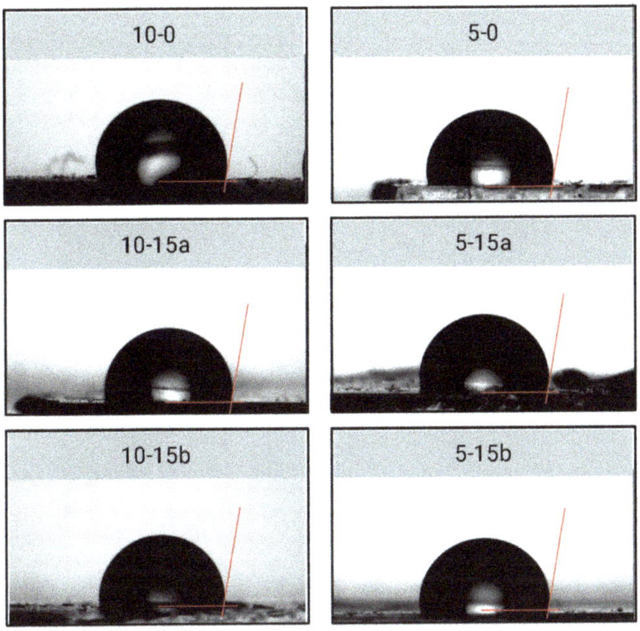

Figure 4. Water droplets on PDMS-SiO_2 selected composites during contact angle testing.

Table 5. Water contact angle and volumetric resistivity results for PDMS-SiO_2 selected composites.

Group	Contact Angle (°)	Resistivity (10^{11} Ω/cm)
15-0	100.4 ± 2.1	3.6 ± 0.6
10-0	99.1 ± 1.6	3.1 ± 0.8
5-0	99.7 ± 5.0	2.5 ± 0.9
10-15a	99.3 ± 2.3	2.6 ± 0.6
5-15b	98.1 ± 3.8	2.4 ± 0.6

No significant change in water contact angle (neither increase nor decrease) was observed in the reinforced composites (10-15a and 5-15b) compared with the unreinforced polymers (10-1 and 5-0). All selected composites exhibited a contact angle greater than 95°, thereby confirming their hydrophobic nature and falling within the normal PDMS contact angle range (95° to 113°) [25]. These results were attributed to the tendency of PDMS to migrate at the surface upon deposition, leading to the formation of a superficial layer predominantly composed of PDMS chains [32].

In addition to their hydrophobic properties, the preservation of the electrical insulation of PDMS coatings is deemed essential for their application in the electrical industry. The conductivity of the composites with superior mechanical performance and higher reinforcement concentrations (10-15a and 5-15b) was evaluated. The volumetric resistivities were determined via conductivity testing using the reversed voltage method [33] and recorded in Table 5.

The volumetric resistivity values obtained for both the reinforced composites (10-15a and 5-15b) and the control groups (10-0 and 5-0) fell within a comparable range, with all samples exhibiting a volumetric resistivity of approximately 10^{11} Ω/cm, without significant differences. These values were consistent with the PDMS reported properties for insulating composites indicating that the addition of SiO_2 reinforcements up to a concentration of 15% neither reduced nor increased the electrical resistivity of these composites [34].

2.2.3. Superficial Adherence

In addition to the surface properties of the composites, maintaining the adhesion of PDMS to ceramic and polymeric substrates is crucial for its application as insulating coatings. Figure 5 presents the results obtained from adhesion tests conducted using the cross-cut tape method, following the ASTM D3359 standard [35].

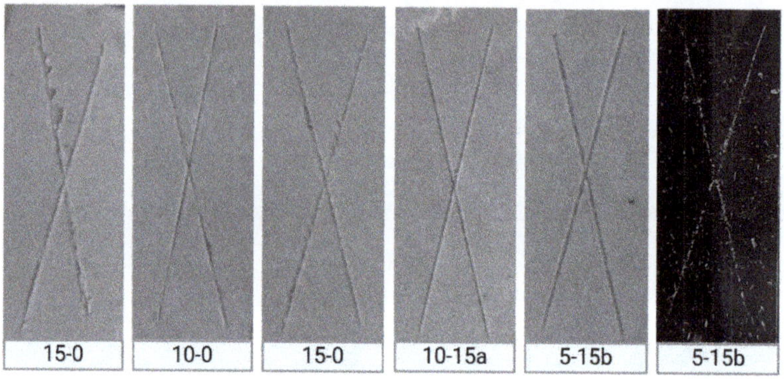

Figure 5. Adhesion test marks following ASTM D3359 on the surfaces of PDMS-%SiO_2 composites.

Composites were analyzed according to the ASTM D3359 standard based on the amount of material removed during testing, with six possible categories (A0–A5). The reinforced composites (10-15a and 5-15b) and their respective control groups (10-0 and 5-0) were classified as A4, indicating the formation of cross-cut marks on the tape after testing.

In contrast, the 15-0 control group was classified as A3, suggesting variation in the adhesion level. This variation appears to be influenced by the crosslinking degree of the PDMS matrix, showing a behavior contrary to that reported for other polymer systems [36].

These classifications are considered indicative of good surface adhesion and align with values reported in previous papers [4]. The adhesion tests demonstrated that the presence of SiO_2 nanoparticles (SiO_2-a and SiO_2-b) within the polymer matrix did not significantly affect the adhesion properties under the ASTM D3359 standard.

3. Discussion

This investigation demonstrated significant changes in the mechanical and thermal behavior of PDMS composites due to modifications in the selected independent variables, including crosslinking agent concentration (TEOS), sol–gel catalyst type, and SiO_2 nanoparticle concentration. These results indicate potential alterations in the chain mobility processes of the polymer, directly influencing the composite's structural characteristics and mechanical performance [37].

During tensile mechanical testing, the crosslinked PDMS chains experience an initial elastic unfolding process (elastic deformation), transitioning from a coiled to an extended spatial conformation. As deformation continued, the rupture of crosslinking bonds (plastic deformation) is initiated [1]. Similarly, during thermal analyses (DMA and TGA), the temperature elevates the vibrational energy of the molecules, progressively enhancing their mobility until the rupture of the crosslinking bonds (~430 °C) occurs, followed by the breakdown of the main polymer chains (~530 °C) [11,28]. These transitions also reflect an improvement in the material's thermal stability, evidenced by elevated degradation of onset temperatures in reinforced systems. Such enhancements in physical properties contrast with the relatively minor changes observed in the chemical structure of the composites, as reported in previous studies [7].

When compared to other PDMS-based composites reported in the literature, the SiO_2-reinforced systems developed in this study exhibited competitive mechanical properties. For instance, PDMS composites reinforced with TiO_2 nanoparticles have reported tensile strengths ranging from approximately 1.6 to 2.8 MPa at loadings between 2.5 and 15 wt%, although presenting a limited elongation capacity [11]. Similarly, CNT-enhanced PDMS composites have achieved tensile strengths between 0.5 and 1.7 MPa at concentrations ranging from 2 to 10 wt%, while largely maintaining the original elongation behavior of PDMS [12]. In contrast, the SiO_2-reinforced composites synthesized herein demonstrated tensile strengths in the range of 2.3–2.6 MPa (Table 1), comparable to TiO_2-based composites, but notably surpassing the deformation capabilities reported for CNT-based composites, all while preserving essential properties such as hydrophobicity and electrical insulation.

The control groups (15-0, 10-0, and 5-0) demonstrated that an increased TEOS concentration led to a significant enhancement in mechanical behavior of the PDMS matrix. This behavior was attributed to an increase in the crosslinking degree within the matrix [30]. As the TEOS concentration increased, the number of crosslinking bonds also rose, restricting the mobility and subsequent sliding of polymer chains. This effect has been particularly evident in mechanical properties such as the elastic modulus [38]. Previous studies conducted by the research group have corroborated this behavior, obtaining a composite with increased stiffness and mechanical strength [4].

Previous investigations have demonstrated that the incorporation of nanostructured reinforcements within a polymeric matrix such as PDMS modifies the internal organization of the polymer chains, resulting not merely in the formation of a two-phase system (matrix–reinforcement) but in a material exhibiting active internal interactions [39]. Unlike micrometer-scale reinforcements, the presence of nanoparticles enables direct interaction between the long polymer chains and the reinforcing particles, thereby altering the material's response to external stimuli, such as mechanical and thermal energy, which are inherently dependent on the molecular dynamics of the system [40].

It is proposed that SiO_2 nanoparticles, like other nanostructured reinforcements, act as high-energy sites within the polymeric matrix [39]. Through electrostatic interactions, primarily Van der Waals forces, these nanoparticles alter the mobility of the polymer chains by physically interacting with the surface hydroxyl (-OH) groups present on the SiO_2 nanoparticles and the methyl groups distributed along the PDMS backbone [41]. Previous studies have demonstrated the absence of new chemical bonds or distinct configurations in PDMS–SiO_2 composites [7], confirming that the interaction is predominantly physical rather than chemical. Since the interaction is electrostatic in nature, complete immobilization of the polymer chains is avoided, unlike what occurs in systems where covalent bonding restricts mobility. Such chemical interactions have been shown to lead to excessive rigidification and a marked reduction in the maximum strain achieved by the material [42]. In contrast, this physical interaction allows the reduction of polymer chain mobility while leading effective load distribution throughout the network, resulting in enhanced thermal and mechanical resistance, without compromising the elastomeric behavior of the PDMS matrix [28,37].

With the above mentioned, the incorporation of SiO_2 nanoparticles with varying sizes and concentrations as reinforcing agents significantly influence the mechanical and thermal performance of the composites. A progressive increase in mechanical properties was observed with increasing nanoparticle content, reaching peak values at the highest reinforcement level (15%) in most of the studied groups. Furthermore, particle properties such as size and agglomeration were shown to directly affect the effective interfacial surface area available for interaction with the polymer chains [43], thereby also influencing the thermal and mechanical performance exhibited by the polymeric composites.

In 15-b and 10-b groups, a decline in ultimate strain and tensile stress was observed at the highest concentrations of reinforcing particles. This behavior has been associated with reinforcement agglomeration processes [20]. The reduction in mechanical performance is attributed to a decrease in the effective interaction area, as the aggregation of reinforcing particles leads to the formation of heterogeneous regions within the material. These regions result in stress concentration points and a subsequent reduction in mechanical strength [4,9].

Among the reinforcement systems used (SiO_2-a and SiO_2-b), the pH of the sol–gel medium directly influenced the final physicochemical properties of the synthesized particles [18]. Previous studies have demonstrated that SiO_2 particles synthesized under alkaline conditions (SiO_2-b) develop larger particle sizes and contain a higher concentration of by-products, such as TEOS, ethanol, and acetone, compared to SiO_2-a [7]. These differences are attributed to distinct reaction pathways followed during the hydrolysis and condensation stages of the sol–gel process [18,19]. These nanoparticles exhibited greater thermal and mechanical properties in composites reinforced with SiO_2-b (Tables 1 and 2).

Although the increased surface area of SiO_2-a particles might initially seem advantageous, the lower surface energy of SiO_2-b particles potentially reduces agglomeration and promotes a higher degree of dispersion [44]. These factors contribute to a more uniform particle size and distribution throughout the composite [45].

Additionally, the higher presence of sol–gel by-products such as TEOS in SiO_2-b [7] influenced the crosslinking conditions of the PDMS system after particle formation. An increased concentration of TEOS directly enhanced the final crosslinking degree, improving the mechanical properties of the polymer [30]. However, at high concentrations, this effect may also lead to excessive rigidity and reduced elasticity [22], as observed in the 15-5b and 10-15b composites.

The mechanical and thermal properties of the composites were enhanced by the formation of a more homogeneous system, which enabled a more uniform distribution of thermal and mechanical energy throughout the polymeric network [11]. Furthermore, an increased crosslinking degree contributed to an improvement in tensile strength [1], with the 5-15b composite exhibiting the most favorable thermal and mechanical performance among the evaluated samples.

The composites exhibiting the highest mechanical performance (5-15b and 10-15a) demonstrated variations in solvent absorption properties and crosslinking efficiency (Table 5). The reduced solvent absorption observed in the composites was attributed to the presence of SiO_2 particles occupying the free spaces between polymer chains, thereby decreasing the intramolecular free space within the structure and consequently limiting the uptake of solvent molecules [46]. Additionally, the minor decrease in crosslinking efficiency (X_y) may result from the higher concentration of unreacted by-products such as ethanol and TEOS, generated during hydrolysis and condensation reactions [19,47]. These small molecules were removed through solvent immersion during the gravimetric test.

The hydrophobicity and volumetric conductivity tests confirmed the absence of significant variations (either increase or decrease) in these essential insulation properties. The preservation of hydrophobic characteristics (>95°) is attributed to the ability of PDMS to form a surface layer upon deposition [5,32]. This phenomenon leads to the formation of a polymer-rich layer, maintaining hydrophobicity across composites with different reinforcer concentration and structure (SiO_2-a and SiO_2-b). The stability of volumetric resistivity in the reinforced composites is associated with the inherently insulating nature of silicon-based structures (SiO_2 NPs and PDMS) that constitute the composites [1,43]. The use of these silicon composite systems ensures the retention of high electrical resistance values (~10^{11} Ω/cm) [24].

Like the insulation properties, adhesion was also maintained upon the addition of reinforcements (SiO_2-a and SiO_2-b), with all composites classified as A4 according to the ASTM D3359 standard [35]. This behavior is also attributed to the formation of an interfacial PDMS layer between the ceramic substrate and the composite [4,5], preventing adhesion from being significantly affected by the reinforcement concentration and type. In contrast, the PDMS:TEOS ratio appeared to induce a distinguishable change in adhesion among the control groups, which was associated with the crosslinking degree of PDMS [48].

The enhanced combination of mechanical reinforcement, hydrophobicity, and electrical insulation properties achieved in the PDMS-SiO_2 composites highlights their potential for broader industrial applications beyond electrical insulation. In the aerospace sector, materials with high elasticity, low density, thermal stability, and environmental resistance are essential for components such as flexible coatings, sealants, and insulation layers in aircraft and spacecraft systems [49]. Likewise, in the biomedical field, reinforced PDMS composites are increasingly explored for the fabrication of polymeric prostheses and flexible implants, where mechanical modification, biocompatibility, and sustained surface hydrophobicity are critical to ensure long-term stability across different types of tissues [50]. Moreover, these composites show significant promise in emerging applications such as wearable electronic devices [51], and microfluidic systems [52], where simultaneous requirements of mechanical flexibility and chemical resistance are critical. Therefore, the development of SiO_2-reinforced PDMS materials offers a versatile platform for advancing next-generation multifunctional devices across diverse technological fields.

4. Materials and Methods

4.1. Materials

For all synthesis and characterization processes conducted in this study, materials were washed with distilled water and dried to remove any contaminants or impurities. The precursors employed for the synthesis of the PDMS-%SiO_2 composites included acetone (C_3H_6O) (99.5%, Meyer®, Mexico City, Mexico), hydroxy-terminated polydimethylsiloxane (PDMS-OH) (2550–3570 cSt, Sigma Aldrich®, Milwaukee, WI, USA), tetraethyl orthosilicate (TEOS) (98%, Sigma Aldrich®, Wuxi, China), ammonium hydroxide (NH_4OH) (28%, J.T. Baker®, Madrid, Spain), nitric acid (HNO_3) (65–70%, J.T. Baker®, Madrid, Spain), and dibutyltin dilaurate (DBTDL) (95%, Sigma Aldrich®, Milwaukee, WI, USA).

4.2. Synthesis of PDMS-SiO_2 Nanocomposites

The in situ synthesis of PDMS-%SiO_2 nanocomposites was conducted using the sol–gel method and the crosslinking reaction of the hydroxyl-terminated PDMS matrix, employing magnetic and ultrasonic stirring techniques, as described in previous investigations [4,7]. The process began with a solution of acetone and PDMS in a weight ratio of 1:0.6 (PDMS:C_3H_6O). The solution was stirred up and homogenized for 1 h. Subsequently, TEOS was introduced according to the desired nanoparticle concentration (0%, 5%, 10%, and 15% by weight). The reported particle concentrations were calculated based on the particle yield obtained during the SiO_2 sol–gel synthesis.

The hydrolysis process and subsequent condensation of SiO_2 nanoparticles were initiated by the dropwise addition of the alkaline or acidic catalyst (NH_4OH or HNO_3), adjusting the pH to 10 (alkaline) or 2 (acidic) [7]. The solution was stirred for 2 h and neutralized to pH 7. TEOS was added in accordance with the selected PDMS:TEOS weight ratios (5:1, 10:1, and 15:1), followed by stirring for 1 h. The solution was degassed in an ultrasonic bath for 15 min to remove any remaining air bubbles and finally stirred mechanically for 5 min.

Crosslinking was initiated by the dropwise addition of the DBTDL catalyst at a weight ratio of 1:0.02 (PDMS:DBTDL). The mixture was mechanically stirred for an additional 5 min, deposited and allowed to be cured at room temperature for 24 h.

4.3. Characterization Techniques

The synthesized composites were mechanically analyzed through tensile tests to select the most suitable study groups for subsequent thermal, electrical, and surface characterizations.

The tensile tests were performed in accordance with ISO 527-1 standards [53], using type 5B dog-bone specimens with five repetitions per group. A Shimadzu AGS-X universal testing machine equipped with a 100 N load cell, a crosshead speed of 30 mm/min, and rough-surface grips were employed. The applied force and displacement were recorded throughout the test until rupture, allowing the determination of the ultimate tensile stress, elastic modulus, and ultimate strain of the composites.

Thermal behavior of the composites was analyzed to evaluate the effect of SiO_2 reinforcers on various energy-related processes within the PDMS matrix. Dynamic mechanical analysis (DMA) was conducted using a Perkin Elmer DMA 7 Dynamic Mechanical Analyzer, employing a heating ramp of 5 °C/min, a static force of 300 mN, a dynamic force of 250 mN, and a temperature range of −130 to 80 °C. Thermogravimetric analysis (TGA) was performed using a thermogravimetric analyzer with a heating ramp of 5 °C/min and a temperature range of 30–700 °C. This thermal characterization provided deeper insights into the nature of the interaction between the polymeric matrix and the reinforcements.

Once the composites with the best mechanical performance were selected, their crosslinking efficiency, hydrophobicity, conductivity, and adhesion properties were assessed to ensure that these intrinsic and surface characteristics were not adversely affected.

Crosslinking efficiency (X_y) was evaluated using gravimetric methods, determining the material loss when dispersed in an ethanol solution by triplicate [54]. Hydrophobicity tests were conducted through contact angle measurements using a Ramé-hart goniometer model 250-U1 with an average water droplet volume of 8.43 µL. Volumetric conductivity was measured using a Keithley 6517B electrometer equipped with an 8009 fixture for high resistivity. Samples with dimensions of 6 × 7 cm and a thickness of 0.7 mm were evaluated using the reverse voltage method for high-resistance measurements, with a sweep range of -250 to 250 V and a current of 20 mA [33]. Finally, surface adhesion was assessed according to the ASTM D3359 standard using the cross-cut tape test method [35]. Material was deposited on a ceramic substrate with an area of 15 cm^2 and an average thickness of 0.5 mm. Each composite was tested in triplicate and categorized on a scale from 0 A to 5 A.

5. Conclusions

The in situ sol–gel synthesis of PDMS-SiO_2 composites enabled the fabrication of advanced insulating materials with significantly enhanced mechanical and thermal performance. The study demonstrated that the concentration of SiO_2 nanoparticles, the pH medium of the sol–gel process (acidic or alkaline), and the PDMS:TEOS ratio are key factors that modify the physicochemical properties of the composites. Among the reinforced polymers, those incorporating alkaline-synthesized SiO_2 (SiO_2-b) exhibited superior mechanical and thermal properties, particularly the 5-15b group, which achieved the highest fracture stress and viscoelastic modulus. This improved behavior was attributed to a more homogeneous size distribution, lower surface energy, and better dispersion of the nanoparticles, which minimized agglomeration and facilitated effective stress transfer within the polymer network. Additionally, maintaining essential surface properties (hydrophobicity, electrical resistivity, and adhesion) confirmed that the in situ incorporation of SiO_2 did not compromise the functional performance of the coatings. These findings supported the implementation of in situ sol–gel synthesis under alkaline catalysis as an effective strategy for the development of PDMS-based nanocomposites with enhanced mechanical and thermal performance for high-demand insulation applications.

Moreover, the in situ approach facilitated the incorporation of nanostructured reinforcements by enabling the identification and control of key experimental variables for the development of next-generation composites. This synthesis strategy provided a versatile platform for the tailored modification of polymeric matrices, thereby broadening the scope of application of these materials both in advanced technological and industrial fields.

Author Contributions: Conceptualization, A.C., R.V.-C. and K.E.; methodology, A.C.; validation, J.V.C.-R., R.V.-C. and K.E.; formal analysis, A.C., J.V.C.-R., R.F.V.-C. and K.E.; investigation, A.C.; resources, J.V.C.-R. and K.E.; data curation, A.C., R.F.V.-C. and K.E.; writing—original draft preparation, A.C.; writing—review and editing, A.C., J.V.C.-R., R.V.-C. and K.E.; supervision, J.V.C.-R., R.V.-C. and K.E.; project administration, K.E. All authors have read and agreed to the published version of the manuscript.

Funding: J.V.C.-R. acknowledges support from SECIHTI Projects 1360 and 248378. K.E. acknowledges support from FONFIVE-2024 (FIN202413).

Institutional Review Board Statement: No applicable.

Informed Consent Statement: No applicable.

Data Availability Statement: Data are contained within the article.

Acknowledgments: A. Cordoba thanks SECIHTI for a granted scholarship. The authors gratefully acknowledge Patricia Quintana at Laboratorio Nacional de Nano y Biomateriales (LANNBIO) Cinvestav-IPN, Unidad Mérida (Project 322758).

Conflicts of Interest: The authors declare no conflicts of interest.

Abbreviations

The following abbreviations are used in this manuscript:

DBTDL	Dibutyltin dilaurate
DMA	Dynamic mechanical analysis
PDMS	Polydimethylsiloxane
TEOS	Tetraethyl orthosilicate
TGA	Thermogravimetric analysis

References

1. Teixeira, I.; Castro, I.; Carvalho, V.; Rodrigues, C.; Souza, A.; Lima, R.; Teixeira, S.; Ribeiro, J. Polydimethylsiloxane Mechanical Properties: A Systematic Review. *AIMS Mater. Sci.* **2021**, *8*, 952–973. [CrossRef]
2. Ghosh, D.; Khastgir, D. Degradation and Stability of Polymeric High-Voltage Insulators and Prediction of Their Service Life through Environmental and Accelerated Aging Processes. *ACS Omega* **2018**, *3*, 11317–11330. [CrossRef] [PubMed]
3. Weththimuni, M.L.; Chobba, M.B.; Sacchi, D.; Messaoud, M.; Licchelli, M. Durable Polymer Coatings: A Comparative Study of PDMS-Based Nanocomposites as Protective Coatings for Stone Materials. *Chemistry* **2022**, *4*, 60–76. [CrossRef]
4. Cordoba, A.; Rivera-Muñoz, E.M.; Velázquez-Castillo, R.; Esquivel, K. PDMS/TiO_2 and PDMS/SiO_2 Nanocomposites: Mechanical Properties' Evaluation for Improved Insulating Coatings. *Nanomaterials* **2023**, *13*, 1699. [CrossRef]
5. Chen, Z.; Su, X.; Wu, W.; Chen, S.; Zhang, X.; Wu, Y.; Xie, H.; Li, K. Superhydrophobic PDMS@TiO_2 Wood for Photocatalytic Degradation and Rapid Oil-Water Separation. *Surf. Coat. Technol.* **2022**, *434*, 128182. [CrossRef]

6. Ain, Q.U.; Wani, M.F.; Sehgal, R.; Singh, M.K. Role of Al_2O_3 Reinforcements in Polymer-Based Nanocomposites for Enhanced Nanomechanical Properties: Time-Dependent Modeling of Creep and Stress Relaxation. *Ceram. Int.* **2024**, *50*, 33817–33838. [CrossRef]
7. Cordoba, A.; Cauich-Rodríguez, J.V.; Vargas-Coronado, R.F.; Velázquez-Castillo, R.; Esquivel, K. A Novel In Situ Sol-Gel Synthesis Method for PDMS Composites Reinforced with Silica Nanoparticles. *Polymers* **2024**, *16*, 1125. [CrossRef]
8. Xia, R.; Zhang, B.; Dong, K.; Yan, Y.; Guan, Z. HD-SiO_2/SiO_2 Sol@PDMS Superhydrophobic Coating with Good Durability and Anti-Corrosion for Protection of Al Sheets. *Materials* **2023**, *16*, 3532. [CrossRef]
9. Tamayo-Vegas, S.; Muhsan, A.; Liu, C.; Tarfaoui, M.; Lafdi, K. The Effect of Agglomeration on the Electrical and Mechanical Properties of Polymer Matrix Nanocomposites Reinforced with Carbon Nanotubes. *Polymers* **2022**, *14*, 1842. [CrossRef]
10. Zhang, Y.F.; Ren, Y.J.; Guo, H.C.; Bai, S. Enhanced Thermal Properties of PDMS Composites Containing Vertically Aligned Graphene Tubes. *Appl. Therm. Eng.* **2019**, *150*, 840–848. [CrossRef]
11. Katz, S.; Lachman, N.; Hafif, N.; Rosh, L.; Pevzner, A.; Lybman, A.; Amitay-Rosen, T.; Nir, I.; Rotter, H. Studying the Physical and Chemical Properties of Polydimethylsiloxane Matrix Reinforced by Nanostructured TiO_2 Supported on Mesoporous Silica. *Polymers* **2022**, *15*, 81. [CrossRef] [PubMed]
12. Du, J.; Wang, L.; Shi, Y.; Zhang, F.; Hu, S.; Liu, P.; Li, A.; Chen, J. Optimized CNT-PDMS Flexible Composite for Attachable Health-Care Device. *Sensors* **2020**, *20*, 4523. [CrossRef] [PubMed]
13. Xu, P.; Jia, Q. Evaluation the Mechanical and Electrochemical Anti-Corrosion Properties of Polydimethylsiloxane/SiO_2 Gel Coated Carbon Steel Rebar in Concrete Pore Solution. *Int. J. Electrochem. Sci.* **2023**, *18*, 100043. [CrossRef]
14. Muhammed Shameem, M.; Sasikanth, S.M.; Annamalai, R.; Ganapathi Raman, R. A Brief Review on Polymer Nanocomposites and Its Applications. *Mater. Today Proc.* **2021**, *45*, 2536–2539. [CrossRef]
15. Kamal, A.; Ashmawy, M.; Sengottaiyan, S.; Algazzar, A.M.; Elsheikh, A.H. Fabrication Techniques of Polymeric Nanocomposites: A Comprehensive Review. *Proc. Inst. Mech. Eng. Part C J. Mech. Eng. Sci.* **2022**, *236*, 4843–4861. [CrossRef]
16. Zheng, C.; Hu, D.; Liu, Z.; Zhang, X.; Yu, K.; Ma, W. A Simple and Efficient in Situ Polymerization of Silica Xerogel-Acrylic Thermal Insulation Coatings. *Prog. Org. Coat.* **2024**, *187*, 108142. [CrossRef]
17. Shibu, M.C.; Benoy, M.D.; Kumar, G.S.; Duraimurugan, J.; Vasudevan, V.; Shkir, M.; AL-Otaibi, O. Hydrothermal-Assisted Synthesis and Characterization of MWCNT/Copper Oxide Nanocomposite for the Photodegradation of Methyl Orange under Direct Sunlight. *Diam. Relat. Mater.* **2023**, *134*, 109778. [CrossRef]
18. Florensa, M.; Llenas, M.; Medina-Gutiérrez, E.; Sandoval, S.; Tobías-Rossell, G. Key Parameters for the Rational Design, Synthesis, and Functionalization of Biocompatible Mesoporous Silica Nanoparticles. *Pharmaceutics* **2022**, *14*, 2703. [CrossRef]
19. Levy, D.; Zayat, M. *The Sol-Gel Handbook: Synthesis, Characterization, and Applications*; Wiley-VCH: Weinheim, Germany, 2015.
20. Demir, E.C.; Benkaddour, A.; Aldrich, D.R.; McDermott, M.T.; Kim, C., II; Ayranci, C. A Predictive Model towards Understanding the Effect of Reinforcement Agglomeration on the Stiffness of Nanocomposites. *J. Compos. Mater.* **2022**, *56*, 1591–1604. [CrossRef]
21. Szymoniak, P.; Pauw, B.R.; Qu, X.; Schönhals, A. Competition of Nanoparticle-Induced Mobilization and Immobilization Effects on Segmental Dynamics of an Epoxy-Based Nanocomposite. *Soft Matter* **2020**, *16*, 5406–5421. [CrossRef]
22. Moreno-Mateos, M.A.; Steinmann, P. Crosslinking Degree Variations Enable Programming and Controlling Soft Fracture via Sideways Cracking. *npj Comput. Mater.* **2024**, *10*, 282. [CrossRef] [PubMed]
23. Syafiq, A.; Vengadaesvaran, B.; Rahim, N.A.; Pandey, A.K.; Bushroa, A.R.; Ramesh, K.; Ramesh, S. Transparent Self-Cleaning Coating of Modified Polydimethylsiloxane (PDMS) for Real Outdoor Application. *Prog. Org. Coat.* **2019**, *131*, 232–239. [CrossRef]
24. Wang, T.-Y.; Mao, J.; Zhang, B.; Zhang, G.-X.; Dang, Z.-M. Polymeric Insulating Materials Characteristics for High-Voltage Applications. *Nat. Rev. Electr. Eng.* **2024**, *1*, 516–528. [CrossRef]
25. Li, S.; Zhang, J.; He, J.; Liu, W.; Wang, Y.; Huang, Z.; Pang, H.; Chen, Y. Functional PDMS Elastomers: Bulk Composites, Surface Engineering, and Precision Fabrication. *Adv. Sci.* **2023**, *10*, 2304506. [CrossRef]
26. Thongchom, C.; Refahati, N.; Roodgar Saffari, P.; Roudgar Saffari, P.; Niyaraki, M.N.; Sirimontree, S.; Keawsawasvong, S. An Experimental Study on the Effect of Nanomaterials and Fibers on the Mechanical Properties of Polymer Composites. *Buildings* **2021**, *12*, 7. [CrossRef]
27. Zhu, L.; Cheng, X.; Su, W.; Zhao, J.; Zhou, C. Molecular Insights into Sequence Distributions and Conformation-Dependent Properties of High-Phenyl Polysiloxanes. *Polymers* **2019**, *11*, 1989. [CrossRef] [PubMed]
28. Bashir, M.A. Use of Dynamic Mechanical Analysis (DMA) for Characterizing Interfacial Interactions in Filled Polymers. *Solids* **2021**, *2*, 108–120. [CrossRef]
29. Deshpande, G.; Rezac, M.E. Kinetic Aspects of the Thermal Degradation of Poly(Dimethyl Siloxane) and Poly(Dimethyl Diphenyl Siloxane). *Polym. Degrad. Stab.* **2002**, *76*, 17–24. [CrossRef]
30. Bardelli, T.; Marano, C.; Briatico Vangosa, F. Polydimethylsiloxane Crosslinking Kinetics: A Systematic Study on Sylgard184 Comparing Rheological and Thermal Approaches. *J. Appl. Polym. Sci.* **2021**, *138*, 51013. [CrossRef]
31. Rajawasam, C.W.H.; Dodo, O.J.; Weerasinghe, M.A.S.N.; Raji, I.O.; Wanasinghe, S.V.; Konkolewicz, D.; De Alwis Watuthanthrige, N. Educational Series: Characterizing Crosslinked Polymer Networks. *Polym. Chem.* **2024**, *15*, 219–247. [CrossRef]

32. Sosnin, I.M.; Vlassov, S.; Dorogin, L.M. Application of Polydimethylsiloxane in Photocatalyst Composite Materials: A Review. *React. Funct. Polym.* **2021**, *158*, 104781. [CrossRef]
33. Lee, H.-G.; Kim, J.-G. Volume and Surface Resistivity Measurement of Insulating Materials Using Guard-Ring Terminal Electrodes. *Energies* **2020**, *13*, 2811. [CrossRef]
34. Shivashankar, H.; Kevin, A.M.; Manohar, S.B.S.; Kulkarni, S.M. Investigation on Dielectric Properties of PDMS Based Nanocomposites. *Phys. B Condens. Matter* **2021**, *602*, 412357. [CrossRef]
35. *ASTM D3359-23*; Standard Test Methods for Rating Adhesion by Tape Test. ASTM International: West Conshohocken, PA, USA, 2017.
36. Lee, J.-H.; Lee, T.-H.; Shim, K.-S.; Park, J.-W.; Kim, H.-J.; Kim, Y.; Jung, S. Effect of Crosslinking Density on Adhesion Performance and Flexibility Properties of Acrylic Pressure Sensitive Adhesives for Flexible Display Applications. *Int. J. Adhes. Adhes.* **2017**, *74*, 137–143. [CrossRef]
37. Li, X.; Li, Z.; Shen, J.; Zheng, Z.; Liu, J. Role of a Nanoparticle Network in Polymer Mechanical Reinforcement: Insights from Molecular Dynamics Simulations. *Phys. Chem. Chem. Phys.* **2021**, *23*, 21797–21807. [CrossRef]
38. Wang, Z.; Volinsky, A.A.; Gallant, N.D. Crosslinking Effect on Polydimethylsiloxane Elastic Modulus Measured by Custom-Built Compression Instrument. *J. Appl. Polym. Sci.* **2014**, *131*, 1–4. [CrossRef]
39. Gupta, N.S.; Lee, K.; Labouriau, A. Tuning Thermal and Mechanical Properties of Polydimethylsiloxane with Carbon Fibers. *Polymers* **2021**, *13*, 1141. [CrossRef]
40. Musa, A.A.; Bello, A.; Adams, S.M.; Onwualu, A.P.; Anye, V.C.; Bello, K.A.; Obianyo, I.I. Nano-Enhanced Polymer Composite Materials: A Review of Current Advancements and Challenges. *Polymers* **2025**, *17*, 893. [CrossRef]
41. Tsige, M.; Soddemann, T.; Rempe, S.B.; Grest, G.S.; Kress, J.D.; Robbins, M.O.; Sides, S.W.; Stevens, M.J.; Webb, E. Interactions and Structure of Poly(Dimethylsiloxane) at Silicon Dioxide Surfaces: Electronic Structure and Molecular Dynamics Studies. *J. Chem. Phys.* **2003**, *118*, 5132–5142. [CrossRef]
42. Ichim, M.; Muresan, E.I.; Codau, E. Natural-Fiber-Reinforced Polymer Composites for Furniture Applications. *Polymers* **2024**, *16*, 3113. [CrossRef]
43. Guo, G.; Zhang, J.; Chen, X.; Zhao, X.; Deng, J.; Zhang, G. Molecular-Dynamics Study on the Thermodynamic Properties of Nano-SiO_2 Particle-Doped Silicone Rubber Composites. *Comput. Mater. Sci.* **2022**, *212*, 111571. [CrossRef]
44. Lazareva, S.V.; Shikina, N.V.; Tatarova, L.E.; Ismagilov, Z.R. Synthesis of High-Purity Silica Nanoparticles by Sol-Gel Method. *Eurasian Chem. J.* **2017**, *19*, 295–302. [CrossRef]
45. He, H.; Fan, G.; Saba, F.; Tan, Z.; Su, Z.; Xiong, D.; Li, Z. Enhanced Distribution and Mechanical Properties of High Content Nanoparticles Reinforced Metal Matrix Composite Prepared by Flake Dispersion. *Compos. Part B Eng.* **2023**, *252*, 110514. [CrossRef]
46. Park, J.; Ha, H.; Yoon, H.W.; Noh, J.; Park, H.B.; Paul, D.R.; Ellison, C.J.; Freeman, B.D. Gas Sorption and Diffusion in Poly(Dimethylsiloxane) (PDMS)/Graphene Oxide (GO) Nanocomposite Membranes. *Polymer* **2021**, *212*, 123185. [CrossRef]
47. Abu Bakar, N.H.; Wan Ismail, W.N.; Mohd Yusop, H.; Mohd Zulkifli, N.F. Synthesis of a Water-Based TEOS–PDMS Sol–Gel Coating for Hydrophobic Cotton and Polyester Fabrics. *New J. Chem.* **2024**, *48*, 933–950. [CrossRef]
48. Sakasegawa, D.; Tsuzuki, T.; Sugizaki, Y.; Goto, M.; Suzuki, A. Effects of Degree of Cross-Links on Adhesion Curves of Cross-Linked Polymers Observed by a Point-Contact Method. *Langmuir* **2010**, *26*, 5856–5863. [CrossRef]
49. Alarifi, I.M. A Comprehensive Review on Advancements of Elastomers for Engineering Applications. *Adv. Ind. Eng. Polym. Res.* **2023**, *6*, 451–464. [CrossRef]
50. Miranda, I.; Souza, A.; Sousa, P.; Ribeiro, J.; Castanheira, E.M.S.; Lima, R.; Minas, G. Properties and Applications of PDMS for Biomedical Engineering: A Review. *J. Funct. Biomater.* **2021**, *13*, 2. [CrossRef]
51. Guan, J.; Fan, W.; Li, H.; Mai, Z.; Jing, Y.; Chen, J.; Zhang, M.; Tang, B.; Yang, Y.; Shen, X. Soft and Stretchable Protective Substrates for Wearable Thermal Managements: Polydimethylsiloxane (PDMS) Composites Containing Paraffin Microcapsules with Silica Nanoshells. *Colloids Surf. A Physicochem. Eng. Asp.* **2024**, *690*, 133809. [CrossRef]
52. Raj, M.K.; Chakraborty, S. PDMS Microfluidics: A Mini Review. *J. Appl. Polym. Sci.* **2020**, *137*, 48958. [CrossRef]
53. *ISO 527-1*; Plastics—Determination of Tensile Properties. ISO: Geneva, Switzerland, 2019.
54. Hirschl, C.; Biebl-Rydlo, M.; Debiasio, M.; Mühleisen, W.; Neumaier, L.; Scherf, W.; Oreski, G.; Eder, G.; Chernev, B.; Schwab, W.; et al. Determining the Degree of Crosslinking of Ethylene Vinyl Acetate Photovoltaic Module Encapsulants—A Comparative Study. *Sol. Energy Mater. Sol. Cells* **2013**, *116*, 203–218. [CrossRef]

Disclaimer/Publisher's Note: The statements, opinions and data contained in all publications are solely those of the individual author(s) and contributor(s) and not of MDPI and/or the editor(s). MDPI and/or the editor(s) disclaim responsibility for any injury to people or property resulting from any ideas, methods, instructions or products referred to in the content.

molecules

MDPI

Article

Sol-Gel Heterogeneization of an Ir(III) Complex for Sustainable Visible-Light Redox Photocatalysis

Janira Herce, Mónica Martínez-Aguirre †, Javier Gómez-Benito, Miguel A. Rodríguez * and Jesús R. Berenguer *

Organometallic Molecular Materials (MATMO), Departamento de Química-Instituto de Investigación en Química (IQUR), Universidad de La Rioja, Madre de Dios, 53, E-26006 Logroño, La Rioja, Spain; janira.herce@unirioja.es (J.H.); m.martinez-aguirre@imperial.ac.uk (M.M.-A.); javiergobe01@gmail.com (J.G.-B.)

* Correspondence: miguelangel.rodriguez@unirioja.es (M.A.R.); jesus.berenguer@unirioja.es (J.R.B.)

† Current address: Department of Chemistry, Imperial College, London, Molecular Sciences Research Hub, White City Campus, Wood Lane, London W12 0BZ, UK.

Abstract: Photocatalysis is a key strategy for the development of sustainable solar-driven chemical processes. In this work, we report the synthesis and characterization of a novel organometallo–ionosilica material derived from the self-condensation of an alcoxysilane functionalized Ir(III) complex. In acetonitrile suspension, the material retains the photophysical properties of its precursor in solution in the same solvent, together with a significant absorption in the visible between 400 and 500 nm. As a heterogeneous photocatalyst, the material showed high efficiency in the reductive dehalogenation of 2-bromoacetophenone under blue light irradiation, achieving high yields of conversion of about 90%, and excellent recyclability in seven catalytic cycles, retaining more than 70% of the catalytic efficiency. All these properties of the self-condensed material highlight its potential as an efficient and sustainable heterogeneous photocatalyst for applications in organic synthesis and solar-driven redox processes.

Keywords: organometallo–ionosilica; visible-light electron transfer photocatalysis; heterogeneous photocatalysis; cyclometalated iridium(III) complex

1. Introduction

In recent decades, growing concerns about the scarcity of fossil fuels and their environmental impact have driven the search for sustainable energy sources. Among these, solar energy has emerged as a promising alternative due to its inexhaustible supply and low environmental footprint. Beyond photovoltaic electricity generation, solar energy can also drive chemical reactions through photocatalysis, a process that, inspired by natural photosynthesis, converts light energy into chemical energy, allowing reactions to occur under milder and more efficient conditions [1]. As a result, photocatalysis has become a key tool in the development of sustainable technologies for solar fuel production, environmental remediation, and the synthesis of high-value compounds [2–4].

The foundations of modern photochemistry were established in the late 19th century by Ciamician and Silber, pioneers in the use of light energy for organic synthesis [5]. Their founding work on photoreduction and photoisomerization remains relevant today and serves as the basis for light-driven chemical transformations [6]. Since then, photocatalysis has evolved significantly, particularly with the development of visible-light-activated systems. Visible light is less energetic than UV, reducing unwanted side reactions and making it a safer and more accessible alternative [7,8]. Recent advances in light-emitting diode

(LED) technology, which provides high-intensity visible light in a controlled-wavelength range, have further expanded photocatalytic applications [9], including organic synthesis, CO_2 reduction, and solar fuel production [10].

Therefore, efficient use of visible light, or even better solar radiation, is desirable and can be effectively achieved using organometallic and coordination complexes of transition metals [11]. Photocatalysts for visible light photocatalysis must meet basic criteria [12], including strong absorption in the visible region, long-lived excited states, and tunable redox properties to facilitate single-electron transfer (SET) or energy transfer (EnT) processes [4]. In this area, cyclometalated Ru(II), Ir(III), and Pt(II) complexes have gained prominence as visible light photocatalysts due to their high photochemical and electrochemical stability [13–15]. Among them, Ir(III) complexes exhibit exceptional photophysical and electronic properties, arising from strong spin-orbit coupling that allows efficient intersystem crossing transitions to long-lived triplet states. These properties, combined with their structural versatility, make them ideal candidates for a wide range of applications, including organic light-emitting diodes (OLEDs) and visible light-driven catalytic processes [16]. The ability to fine-tune their redox and photophysical properties through ligand design further enhances their versatility in synthetic and applied chemistry [17].

Despite their advantages, Ir(III) complexes have one major limitation: their use in homogeneous catalysis complicates their recovery and reuse, increasing costs and raising environmental concerns [13,14,18]. A promising strategy to overcome this problem is the heterogeneization of these complexes by using suitable inorganic substrates, such as metal nanoparticles, MOFs, zeolites, or inorganic oxides like silica [7,19,20]. This strategy has already been successfully applied to several areas of current research interest, including water splitting [21], solar fuel production [22,23], and chemical synthesis [10,24]. In this context, our groups have recently described the in situ preparation of several hybrid organometallo–silicas based on the iridium cyclometalated complex [Ir(dfppy)$_2$(dasipy)]PF_6 (dfppy = 2-(2,4)-difluorophenylpyridinyl; dasipy = N,N′-dipropyltriethoxysilane-2,2′-bipyridine-4,4′-dicarboxamide) and their application and good recyclability in the energy transfer (EnT) photoisomerization reaction of *trans*- to *cis*-stilbene under blue illumination and mild conditions [25]. Nevertheless, these materials showed low or no efficiency for single electron transfer (SET) processes, probably due to the insulating nature of silica. To overcome this issue, we have recently reported a novel material obtained by the self-condensation of the terminal alcoxysilane groups of the complex [Ir(dfppy)$_2$(dasipy)]PF_6 [26]. This material consists of discrete nanoparticles formed exclusively by the organometallic cations linked by Si–O bonds, and the anionic PF_6^- counterparts. Thus, this organometallo–ionosilica has to be considered as a new type of ionosilica [27–29], a class of materials with potential applications in energy storage, catalysis, and biomedicine. In fact, it is highly efficient in photoredox reactions under blue light in both oxidative and reductive pathways, with excellent recyclability [26].

All these nanostructured heterogeneous photocatalysts [25,26] have been designed starting from the highly emissive derivative [Ir(dfppy)$_2$(dasipy)]PF_6. Similar to other heteroleptic cyclometallated Ir(III) complexes, in this derivative the LUMO, where the electronic charge is concentrated upon excitation of the molecule, is mainly located on the diimine ligand (dasipy, N^N), which is functionalised with the terminal alkoxysilane groups. The objective of this work is therefore twofold: firstly, to achieve the steric liberation of the diimine ligand from these functional groups and, secondly, to simultaneously increase the electronic delocalisation of this ligand. The alkoxysilane groups should therefore be placed on the cyclometallated ligands (C^N), which are usually less susceptible to functionalisation. In this paper we describe the synthesis of the complex [Ir(Si-dfbzapy)$_2$(pyraphen)]PF_6 (**2**), which fulfils both requirements, and its self-condensation reaction to give the new organometallo–ionosilica **SC-2**. We have also evaluated the photocatalytic activity and reusability of this material under heterogeneous conditions in a photoredox reaction. Specifically, we chose the reductive dehalogenation of 2-bromoacetophenone, as this reaction has been widely used as a benchmark for assessing photocatalyst performance under both homogeneous and heterogeneous conditions [30,31]. The reaction yields acetophenone, a non-toxic precursor in organic synthesis that is widely used in cosmetics, pharmaceuticals, and resins [32]. Through comprehensive characterisation and catalytic testing of **SC-2**, this study aims to contribute to the development of sustainable photocatalytic systems with potential applications in energy conversion and environmental sciences.

2. Results

2.1. *Synthesis and Characterization of Complexes* **1** *and* **2** *and the Self-Condensed Ionosilica* **SC-2**

With the aim of obtaining a final Ir(III) complex with cyclometalated ligands functionalized with alcoxysilane groups, the HC^N compound 2,6-difluoro-3-(pyridine-2-yl) benzaldehyde (dfbzapyH) [33] was used to obtain the organometallic chloride-bridged $[Ir(dfbzapy)_2(\mu\text{-}Cl)]_2$ and the cationic solvate $[Ir(dfbzapy)_2(NCMe)_2]PF_6$ precursors, following the procedure described in the bibliography [34]. The reaction of the solvate precursor with the diimine ligand pyrazino[2,3-f][1,10]phenanthroline (pyraphen) in refluxed dichloromethane for 6 h afforded the cationic complex $[Ir(dfbzapy)_2(pyraphen)]PF_6$ (**1**, see Section 3 for more details, and Scheme S1 in SI). The imination reaction of the aldehyde fragments of complex **1** with 3-aminopropyltriethoxysilane (APTES) in THF for 1 h yielded complex $[Ir(Si\text{-}dfbzapy)_2(pyraphen)]PF_6$ (**2**), which is functionalized with terminal alcoxysilane groups on both of the cyclometalated ligands (see SI, Scheme S2). Due to its high tendency to self-condense in the solid state, preventing further solubility, complex **2** was not isolated in solid state and was therefore handled in THF solution. Complexes **1** and **2** were characterized by high resolution mass spectrometry (ESI) and multinuclear ^{1}H, $^{13}C\{^{1}H\}$ and ^{19}F NMR spectroscopy. In both cases, NMR spectra show the presence of only one diimine pyraphen ligand and the equivalence of both cyclometalated groups (see Section 3, and Figures S1 and S2 in SI). In addition, complex **1** was characterized in the solid state by X-ray crystallography (Figure 1, see also Tables S1 and S2 in the SI). The molecular cation ([1]+) presents the expected pseudo-octahedral environment with the nitrogen atoms of both cyclometallated ligands in a mutually *trans* disposition, which is also the geometry expected for complex **2** and the molecules integrating the material **SC-2**. The bond distances and angles (see SI, Table S1) are similar to those observed for other related cationic heteroleptic $[Ir(CN)_2(NN)]^+$ derivatives [34]. The complex crystallizes in the centrosymmetric $C_{2/c}$ space group with both enantiomers (Δ and Λ) present in the unit cell. The synthesis of the self-condensed material **SC-2** was achieved by controlled addition of the solution of complex **2** in THF over water. Stirring the mixture in the presence of NaF under reflux for 96 h afforded the ionic material **SC-2** as a yellow solid (Scheme 1, see Section 3 for more details). This material consists exclusively of cationic molecules of complex **2**, which are covalently linked through the condensation of the terminal alkoxysilane groups, together with the corresponding PF_6^- anions.

Figure 2 shows the IR spectra of complexes **1**, **2** and the ionosilica **SC-2**, showing in all the cases the appearance of a P-F stretching vibration at ca. 800 cm^{-1}, indicative of the presence of PF_6^- anions. According to the imination process of the aldehyde groups of complex **1** (SI, Scheme S2), there are several differences between the spectra of complexes **1** and **2**. For instance, the disappearance of the ν(C=O) band at 1963 cm^{-1} observed for complex **1** and the appearance in the spectrum of complex **2** of more intense signals in the high-energy region (around 2950 cm^{-1}), attributed to the C–H stretching vibrations of aliphatic groups, as well as two very strong absorptions corresponding to the stretching vibrations of Si–O–C (1071 cm^{-1}) and Si–O (1006 cm^{-1}) bonds. As expected, the IR spectrum of the self-condensed material **SC-2** is quite similar to that observed for complex **2**, indicating that the cyclometalated Ir(III) cations retain their structure during the condensation process. Analogous to the synthesis of a related ionosilica based on complex $[Ir(dfppy)_2(dasipy)]PF_6$, recently reported by our group [26], the main difference between the spectra of complex **2** and material **SC-2** is the significant weakness of the absorption bands related to Si–O–C bonds in the material. Since **SC-2** consists of self-condensed molecules of complex **2**, each containing six ethoxide groups susceptible to hydrolysis and condensation, it is highly probable that not all these groups undergo complete hydrolysis to form Si–O–Si bonds. Indeed, these reactions may not occur to the same extent in all molecules. Scheme 1 therefore represents an average molecular model of this material. Under these conditions, the hydrolysis of two ethoxide groups per triethoxysilane moiety is assumed, resulting in a material with the empirical formula $C_{48}H_{42}F_{10}IrN_8O_4Si_2P$, considering the presence of the PF_6^- anion. This empirical formula agrees with the elemental analysis results, which consistently give values corresponding to the presence of eight water molecules per cationic unit (see Experimental Section in SI). Finally, the preservation of the structural integrity of the cationic organometallic fragments is in agreement with the MALDI(+) spectrum in the solid state, which shows a signal at m/z 1065 consistent with the isotopic distribution of the ionization fragment $[\textbf{SC-2} - \{(CH_2)_3Si(OEt)_3\} + 3H]^+$. It should be noted that the MALDI(−) spectrum also shows a signal at m/z 145 corresponding to the presence of the PF_6^- anion (see SI, Figure S3).

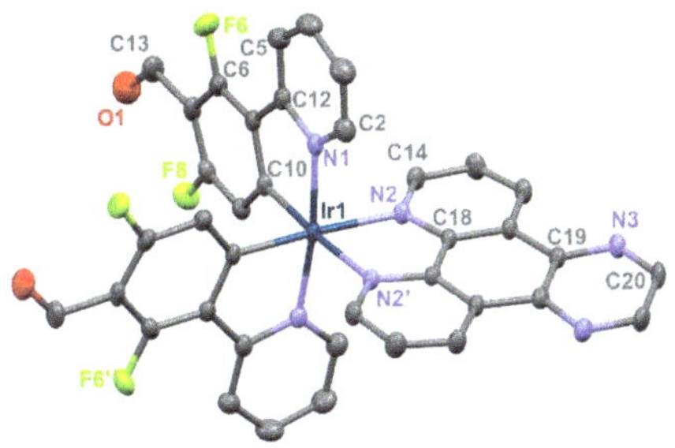

Figure 1. Molecular structure of the complex cation $[Ir(dfbzapy)_2(pyraphen)]^+$ ($\mathbf{1^+}$). Most relevant bond lengths (Å): Ir-C 2.006(4); Ir-N(1) 2.044(4); Ir-N(2) 2.141(2).

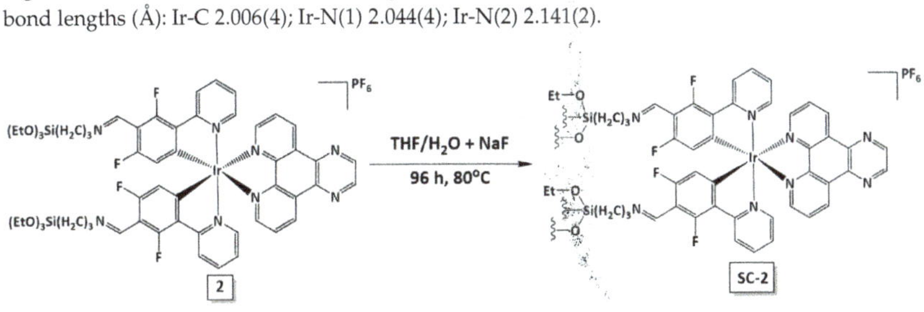

Scheme 1. Synthesis of the organometallo–ionosilica **SC-2**, along with its proposed average composition.

The TEM images (Figure 3; see also Figure S4 in SI) show that **SC-2** consists of aggregates with sizes ranging between 200 and 1000 nm, formed by the agglomeration of smaller nanoparticles. The FESEM analysis also shows the granular texture typical of silica-based materials obtained by hydrolysis of alcoxysilane groups [35], while the electron diffraction pattern of **SC-2** confirms the amorphous nature of the material (see SI, Figure S4). Elemental mapping of **SC-2** using FESEM-EDX analysis (Figure 3) reveals a homogeneous distribution of atoms over the surface of the material, with a uniform composition throughout the structure. Notably, phosphorus atoms are exclusively associated with the anion PF_6^-, confirming their integral role in the material. These anions are likely to be located not only on the surface but also within the structure to maintain ionic neutrality.

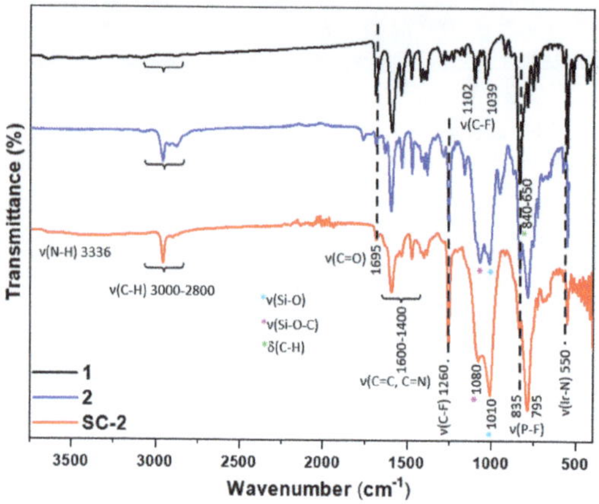

Figure 2. FTIR spectrum of complexes **1**, **2** and the self-condensed material **SC-2**.

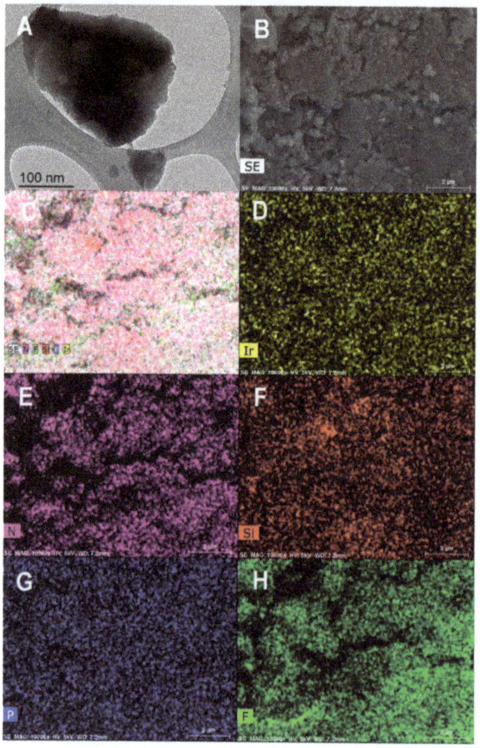

Figure 3. Representative TEM ((**A**), scale bar = 100 nm) and FESEM ((**B**), scale bar = 2 μm) images of the self-condensed material **SC-2**, with the corresponding layered image (**C**) combining the FESEM-EDX elemental mapping images of Ir (**D**), N (**E**), Si (**F**), P (**G**), and F (**H**) (scale bar in C–H = 2 μm).

2.2. Photophysical and Electronic Characterization

Tables 1 and 2 and Figures 4 and 5 (see also SI, Figure S5) show the absorption and emission data of complexes **1**, **2** and material **SC-2** at room temperature. Figures S5 and S6 and Tables S3 and S4 (see SI) also present the electrochemical properties of complexes **1**, **2** and material **SC-2** at room temperature. The spectra of complex **2** has not been recorded in the solid state due to its high tendency to self-condense in that state. Solution or suspension studies were carried out in acetonitrile, as this is the solvent in which the photocatalytic studies were carried out (see below).

Table 1. Absorption data of complexes **1**, **2** and self-condensed material **SC-2**.

Sample	Medium	λ_{abs}/nm ($\varepsilon \cdot 10^{-3}$/$M^{-1} \cdot cm^{-1}$)
[Ir(dfbzapy)$_2$(pyraphen)]PF$_6$ (1)	Solid	245, 290, 341, 380, 440
	MeCN 5×10^{-5} M	255 (20.0), 296 (11.0), 355 (2.2), 385 (0.9), 442 (0.1)
[Ir(Si-dfppy)$_2$(pyraphen)]PF$_6$ (2)	MeCN 5×10^{-5} M	256 (20.0), 283 (15.4), 360 (2.3), 390 (0.9), 442 (0.1)
SC-2	Solid	256, 295, 350, 375, 420, 445

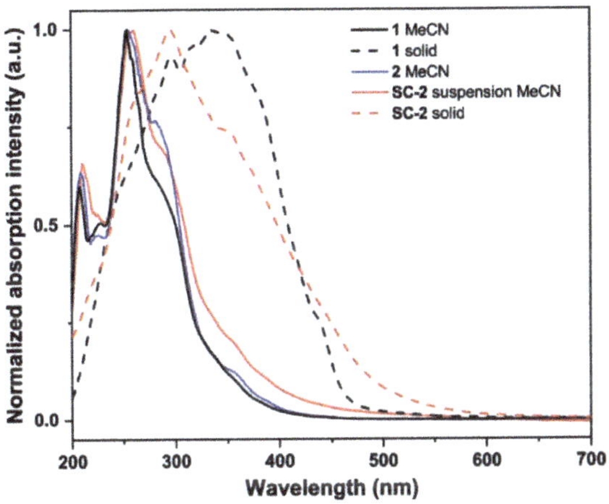

Figure 4. Normalized absorption spectra of complex **1** and **SC-2** in solid state, complex **1** and **2** in solution (MeCN, 5×10^{-5} M), and **SC-2** in suspension of MeCN (1 mg/mL).

As observed in Figure 4, the spectra of complexes **1** and **2** in acetonitrile solution, as well as that of the **SC-2** material in suspension in the same solvent, are very similar, and have been assigned according to the spectra observed in the literature for other related cyclometalated Ir(III) derivatives [26] and the theoretical calculations carried out for complex **2** in MeCN solution (see SI, Figures S7–S9 and Tables S5–S7). Both spectra show the higher-intensity bands between 250 and 300 nm that can be mainly ascribed to spin-allowed intraligand (^{1}IL or ^{1}IL′; L = C^N, L′ = N^N) π-π^* transitions with a certain metal contribution. Between 300 and 400 nm, the spectra show less intense allowed features related with mixture of intraligand transitions and ligand-to-ligand (^{1}LL′CT; C^N→N^N) or metal-to-ligand (^{1}ML′CT; Ir→N^N) charge transfer transitions. Finally, at lower energies (ca. 440 nm), very weak absorptions appear, more intense for the suspension of **SC-2** in MeCN, mainly related to the HOMO-LUMO gap. As observed in the theoretical calculations carried out on complex **2** in MeCN solution (see SI, Table S5 and Figure S7), the HOMO is mainly located on the iridium atom (31%) and the phenyl rings of the cyclometalated ligands (68%), while the LUMO is centred on the diimine ligand (98%). Thus, these low-energy absorptions are attributed to a mixture of ligand-to-ligand and metal-to-ligand charge transfers, traditionally ascribed to spin-forbidden excitations (^{3}ML′CT/^{3}LLCT), due to the strong coupling (SOC) associated with iridium [36]. As expected, the profile observed for the solid state DRUV (Diffuse Reflectance UV-vis, Figure 4) of **1** and **SC-2** appear to be quite different. Nevertheless, the spectra show similar absorption maxima to those of the solution spectra (Table 1). Finally, it should be noted that the low energy absorption observed for **SC-2** at 440 nm in the solid state display a long tail up to about 600 nm.

As observed in Figure 5 (and Table 2), the emission spectra of complexes **1** and **2** in MeCN solution, as well as the spectra of a suspension of **SC-2** in the same solvent, are similar, showing an unstructured green phosphorescence with long lifetimes (τ 12–20 μs) and low quantum yields (ϕ 1.2–7%). The SOMO and SOMO-1 orbitals calculated for complex **2** (see SI, Figure S9) show a similar composition to that described for the LUMO and HOMO (see SI, Figure S7), so this emission can be attributed to a mixture of ^{3}ML′CT/^{3}LL′CT excited states, with a remarkable metallic contribution. For complex **2**, the emission maximum (λ_{em} 500 nm) is slightly blue-shifted with respect to that of complex **1** (λ_{em} 510 nm), probably due to the change in the phenyl ring of the cyclometalated ligands from an aldehyde group in complex **1** to a more donor imine group in complex **2**. However, the emission maximum of the suspension of **SC-2** in MeCN is again slightly shifted to 510 nm. The solid-state spectrum of complex **1** shows a less intense phosphorescence (τ 13.24 μs, ϕ 1%) with a broader profile and a bathochromic shift ($\Delta\lambda_{em}$ 50 nm), although this yellow emission can still be mainly attributed to the mixture of ^{3}ML′CT/^{3}LL′CT excited states. More surprising is the large shift observed in the spectrum of the solid **SC-2** material, which shows a low intensity and broad red phosphorescence (λ_{em} 660 nm, τ 10.73 μs, ϕ 1%). This emission could be due to heterogeneity effects of the amorphous material and/or to the formation of excimers absents in suspension due to the interaction of the molecular cations located on the surface of the material with the polar solvent. The energy of the first triplet state (T_1) and the optical gap between the ground state (S_0) and the T_1 ($E_{0,0}(T_1)$) of complexes **1** and **2** (solution) and material **SC-2** (suspension) have been calculated from the excitation and emission spectra (Figures 5 and S5, Table S3). In all the cases, the experimental T_1 energy values are practically identical (ca. 2.83 eV) and agree well with that calculated theoretically by DFT for complex **2** (2.74 eV), while the suspension of **SC-2** shows the larger value of the optical gap (2.85 eV for **SC-2** vs. 2.78 eV for **1** and **2**). As a model for the electrochemical properties of the material **SC-2** in acetonitrile suspensions, the determination by cyclic voltammetry (CV) of the ground-state potential of complex **2** in MeCN solutions ($E_{ox}{}^{1/2}$ 1.72 V, $E_{red}{}^{1/2}$ −1.61 V. See SI, Figure S6 and Table S4) gives an estimated HOMO-LUMO gap of 3.33 V for this complex, which is also in agreement with the value obtained by theoretical calculations (TD-DFT 3.35 V). Finally, all these data allow the determination of the excited-state redox potential of complex **2** in MeCN (E^*_{red} 1.17 V, E^*_{ox} −1.06 V, Table S4) [37].

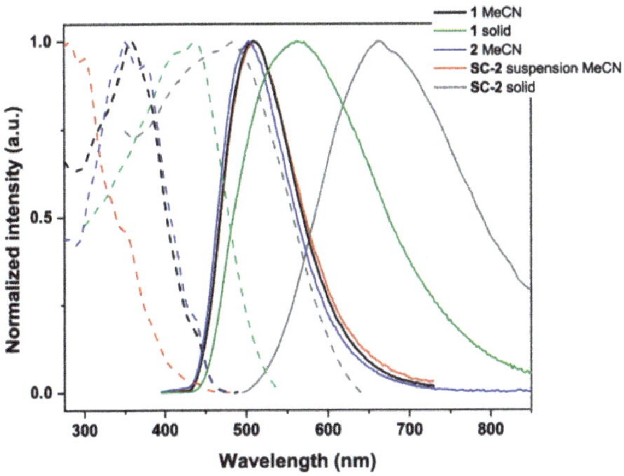

Figure 5. Normalized excitation (dotted line) and emission (solid line) spectra of **1** and **SC-2** in solid state, **1** and **2** in solution (MeCN, 5×10^{-4} M), and **SC-2** in suspension of MeCN (1 mg/mL) at room temperature.

Table 2. Photophysical data in solid state for complex **1**, and **SC-2** in solid state, complex **1** and **2** in solution (MeCN, 5×10^{-4} M), and **SC-2** in suspension of MeCN (1 mg/mL). All data at room temperature.

Sample	Medium	λ_{em}/nm (λ_{exc}/nm)	$\tau/\mu s$ [1]	Φ/%
$[Ir(dfbzapy)_2(pirafen)]PF_6$ (1)	Solid	560 (440)	13.24	1
	MeCN	510 (360)	12.62	7
$[Ir(Si\text{-}dfppy)_2(pirafen)]PF_6$ (2)	MeCN	500 (375)	20.19	1.2
SC-2	Solid	660 (475)	10.73	1
	Suspension MeCN	510 (360)	13.79	2.4

[1] Emissions lifetimes calculated as mono-exponential decay or an average of a bi-exponential decay.

2.3. Photocatalytic Studies

As the organometallo–ionosilica **SC-2** exhibits absorption and excitation maxima at λ_{abs} > 400 nm in acetonitrile suspensions (Figures 4 and 5), its photocatalytic activity has been evaluated under blue light illumination (50 W blue LED, λ_{max} = 450 nm). The reductive dehalogenation reaction of 2-bromoacetophenone was chosen for this purpose (Scheme 2, up). This single electron transfer (SET) reaction requires the presence of three equivalents of triethanolamine (TEOA) as a sacrificial reagent. The reaction starts with the excitation of the photocatalyst, resulting in a spin-allowed metal-to-ligand charge transfer (^{1}MLCT), followed by intersystem crossing from the singlet excited state (^{1}MLCT) to a lower-energy triplet state (^{3}MLCT). The excited iridium species reacts with triethanolamine, producing a reduced iridium radical and a radical cation derived from triethanolamine. The iridium radical anion then returns to its ground state by transferring an electron to 2-bromoacetophenone, thereby facilitating the cleavage of the C-Br bond. This process results in the formation of a bromide anion and an acetophenone radical. The latter abstracts a hydrogen atom from the triethanolamine radical cation, yielding an iminium cation and acetophenone (Scheme 2, down). The reaction progress was monitored by ^{1}H NMR spectroscopy, using 1,3,5-trimethoxybenzene as an internal standard (see SI, Figure S10). Initially, the reaction with the **SC-2** photocatalyst was carried out under N_2 atmosphere on an NMR scale, using deuterated acetonitrile as the solvent, starting from 15 mg of 2-bromoacetophenone and a catalyst loading of 1 mol% (Table 3). Under these conditions, the reaction gives about 90% yield of acetophenone within 9 minutes, but with the complete consumption of the starting bromoacetophenone. This makes **SC-2** a more efficient photocatalyst than the related iridium-based self-condensed material recently reported by us [26], which achieves a similar 90% yield but in 90 min under the same conditions. In addition, the complete consumption of bromoacetophenone indicates the occurrence of a secondary reaction causing the decomposition of approximately 10% of the starting material. Notably, no additional aromatic signals were observed in the final NMR spectra. This behaviour could be explained by the greater electronegativity of bromine compared to carbon, which withdraws electron density from the C-Br bond, thereby increasing the positive charge density of the carbonyl group. As a result, 2-bromoacetophenone would become more reactive towards nucleophilic species present in the reaction medium, starting the decomposition reaction. Also, in the absence of blue light irradiation or TEOA, no reaction was observed in the control reactions (Table 3).

Table 3. Results of the dehalogenation reaction of 2-bromoacetophenone in heterogeneous conditions under N_2, using the organometallo–ionosilica **SC-2** (1 mol% [Ir]) as photocatalyst.

Entry	Time (min)	Yield (%)
1	9	93 (±4) [1]
2 [2]	20	-
3 [3]	20	-

[1] Average value of three runs. [2] No TEOA. [3] No light.

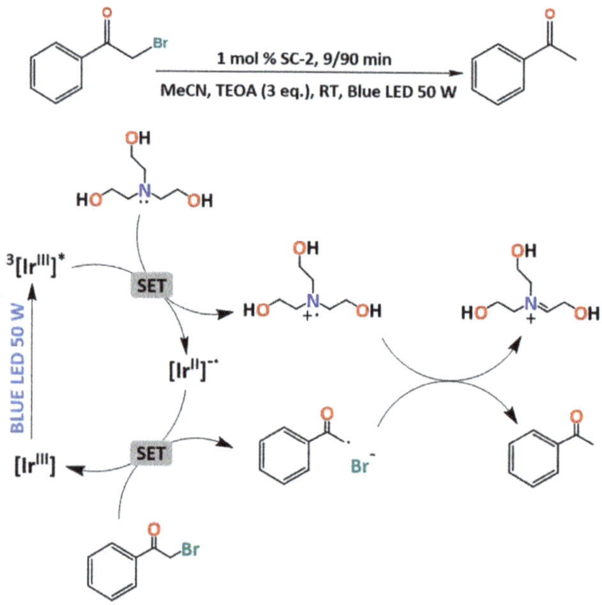

Scheme 2. Photocatalytic dehalogenation of 2-bromoacetophenone (**up**) and mechanism of the reaction (**down**).

To evaluate the recyclability of the heterogeneous photocatalyst and to facilitate its recovery, the process was scaled up while maintaining the same light source under N_2 atmosphere. The first cycle was started with 36.5 mg of **SC-2** (1 mol% catalyst loading), using 0.65 g of 2-bromoacetophenone, 0.30 g of trimethoxybenzene, and 1.3 mL of triethanolamine in each cycle. After each cycle, the photocatalyst was recovered by centrifugation. The material was first washed with water to remove the iminium salt formed during the reaction. It was then washed with absolute ethanol and acetonitrile before being dried and reused in the next catalytic cycle (see Section 3 for more details).

As can be seen in Figure 6, under these conditions, the reaction of the first cycle reaches approximately 60% conversion after 90 min, while the maximum conversion obtained in the last cycle, in the same time period, is 41.5%. This indicates a loss of the catalytic capacity of the **SC-2** material of less than 30% over the seven cycles recorded. Furthermore, two significant decreases were observed (about 9% after the first and sixth cycles), while the performance remained practically stable in the rest of the cycles. This suggests that some photocatalyst may have been lost during the washing processes. This finding is consistent with the observation that no signals corresponding or compatible with the presence of compound **2** have been detected by RMN spectroscopy in the reaction medium of any of the cycles, suggesting that no significant leaching has occurred during the reaction.

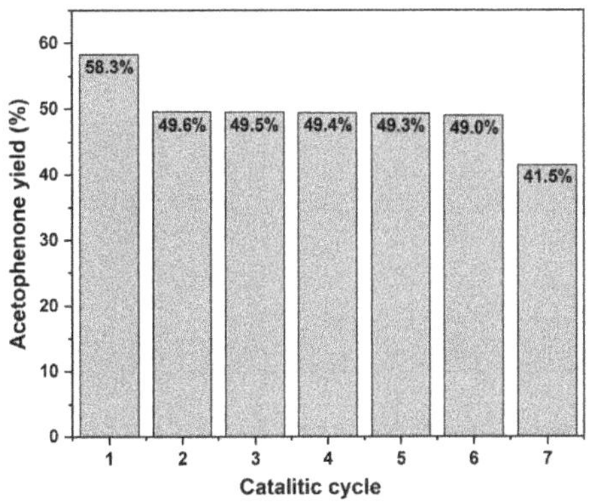

Figure 6. Results of the conversion values from the recovery and reuse of **SC-2** over seven consecutive catalytic cycles, each lasting 90 min. All experiments were conducted under identical conditions.

Finally, to further study the possible photodegradation of **SC-2**, the material used in up to seven catalytic cycles was recovered and studied by means of infrared, absorption, and emission spectroscopies (see SI, Figures S11–S13 and Tables S8 and S9). The IR spectrum of the reused **SC-2** material (Figure S11) is quite similar to that of the original material, with the most significant profile modification occurring in the region corresponding to the absorptions derived from the aromatic groups (1300–1600 cm^{-1}). The DRUV spectra of both materials (Figure S12, Table S8) shows the largest changes. The absorptions at lower energies (λ_{abs} 350–450 nm; ML′CT/LL′CT) are less intense in the spectrum of the reused material, although they appear at similar wavelengths than those observed for the original material. However, the profile of the absorptions between 250 and 300 nm shows a greater change with an increase in the intensity of the bands at ca. 260 nm. These modifications in the high-energy bands have previously been associated with structural changes in the aromatic ligands [37], which would be consistent with the feature observed in the infrared spectrum, indicating some photodegradation of the molecular cations located onto the surface of the material. However, despite the aforementioned, the emission spectra of both materials (Figure S13, Table S9) show similar unstructured red phosporescence at 660 nm, with a small decrease in lifetime (10.7 vs. 13.7 μs), indicating that the bulk of the material exhibits adequate photostability after seven cycles of use.

3. Materials and Methods

3.1. General Methods

The compound 2,6-difluoro-3-(pyridine-2-yl) benzaldehyde (dfbzapyH) [33] and the precursors $[Ir(dfbzapy)_2(\mu\text{-}Cl)]_2$ and $[Ir(dfbzapy)_2(NCMe)_2]PF_6$ were synthesized by procedures described in the bibliography [34]. The ligand pyraphen = pyrazino[2,3-f][1,10]phenanthroline is commercially available. All synthetic procedures for complexes 1 and 2 were performed under an inert atmosphere and in anhydrous conditions. Commercially sourced reagents were used as received, without additional purification. Solvents were obtained from a solvent purification system (M-BRAUN MS SPS-800). Elemental analyses of the compounds were recorded on a Thermo Finnigan Flash 1112 microanalyzer (Thermo Fischer Scientific, Waltham, MA, USA). Mass spectra were made on a Microflex MALDI-TOF Bruker (MALDI) (Bruker, Billerica, MA, USA) spectrometer, operating in the linear and reflector modes using dithranol as matrix, and on a Bruker Microtof-Q spectrophotometer (Bruker, Billerica, MA, USA) with ESI/APCI interface in positive ion mode, using a 90:10 methanol:water mixture and 0.1% formic acid as mobile phase. NMR spectra of the compounds were obtained on Bruker ARX-300 and ARX-400 spectrometers (Bruker, Billerica, MA, USA) using CH_3CN and acetone-d_6 as solvent. Chemical shifts are presented in parts per million (ppm) relative to external standards ($SiMe_4$ for ^1H and $^{13}C\{^1H\}$ and $CFCl_3$ for $^{19}F\{^1H\}$) and coupling constants in hertz (Hz). ^1H-^1H COSY, ^1H-^{13}C HMBC and ^1H-^{13}C HSQC correlation experiments have been used for the assignation of the ^1H and $^{13}C\{^1H\}$ NMR spectra. IR spectra were recorded on a Perkin Elmer FT-IR Spectrometer Spectrum Two in the wavenumber range from 4000 to 400 cm^{-1}. The UV-VIS absorption spectra in solution were obtained using a Hewlett Packard 8453 spectrophotometer (Agilent Technologies, Santa Clara, CA, USA). Solid-state Diffuse Reflectance UV-VIS (DRUV) spectra were recorded on a Shimazdu UV-3600 spectrophotometer (Shimadzu Corporation, Kyoto, Japan) with a Harrick Praying Mantis accessory and recalculated using the Kubelka Munk function. The samples were diluted in KBr powder for the complex and silica for the materials. Excitation and emission spectra were measured on an Edinburgh FLS 1000 spectrofluorimeter (Edinburg Instruments Ltd., Edinburg, UK), where lifetime measurements were also performed with a μF2 pulse lamp (Power: 100 W, Fuse: 3.15 Amp A/S). Quantum yields were measured with Hamamatsu Absolute PL Quantum Yield Measurement System (Hamamatsu Photonics, Shizuoka, Japan). Cyclic voltammetry and differential pulse voltammetry measurements were carried out on a Voltalab PST050 electrochemical workstation (Volta Labs, Boston, MA, USA), using a 0.1 M solution of tetrabutylammonium hexafluorophosphate (NBu_4PF_6) as the supporting electrolyte. A conventional three-electrode configuration was used, consisting of a platinum spar working electrode, a platinum wire counter electrode, and an Ag/AgCl reference electrode. As the solvent, was used CH3CN ([1] 10^{-4} M) under inert atmosphere and the ferrocene/ferrocenium couple served as internal reference.

The morphology of the materials was investigated by transmission electron microscopy (TEM) and Scanning Electron Microscopy (SEM). Samples were prepared by dipping a sonicated suspension of the sample in ethanol on a carbon-coated copper. TEM images were performed using a JEM-2010 microscope (JEOL, 200 kV, 0.14 nm of resolution) (JEOL Ltd., Tokyo, Japan), with a detector of Si(Li) (area of 30 mm^2 and resolution of 142 eV). The digital analysis of the TEM micrographs was performed using DigitalMicrographTM 3.6.1. by Gatan. SEM analyses were carried out in a field emission scanning electron microscope (FESEM) Merlin VP Compact (Zeiss, 1.6 nm of resolution at 1 kV) (Zeiss, Wetzlar, Germany), where EDX analyses were also performed.

3.2. Synthesis of [Ir(Dfbzapy)$_2$(pyraphen)]PF$_6$ (1)

0.83 g (0.36 mmol) of pyrazino[2,3-f][1,10]phenanthroline (pyraphen) were added to a solution of 0.37 g (0.46 mmol) of [Ir(dfbzapy)$_2$(NCMe)$_2$]PF$_6$ in 25 mL of dichloromethane, and the mixture was refluxed for 6 h in the absence of light. After the reaction time, the mixture was evaporated to dryness, and the solid residue was treated with diethyl ether and hexane to yield complex **1** as a yellow solid (0.35 g, 81%). Anal. Calc. for $C_{38}H_{20}F_{10}IrN_6O_2P$: C, 45.38; H, 2.00; N, 8.36. Found: C, 45.38; H, 2.21; N, 9.70.ESI (+): m/z 861 [Ir(dfbzapy)$_2$(pyraphen)]$^+$ (100%). IR (cm^{-1}): ν(C-H$_{aromatic}$) 3096 (w); ν(C=O) 1693 (m); ν(C=C, C=N) 1595 (s), ν(C-F) 1102 (m), 1039 (s); ν(P-F) 837 (vs); ν(Ir-N) 557 (s). ^{1}H NMR (400 MHz, acetone-d$_6$, δ): 10.30 (s, 2H, CHO); 9.83 (dd, $J_{H\text{-}H}$ = 9.0, 1.6 Hz, 2H, H$^{2'}$); 9.36, (s, 2H, H$^{8'}$); 8.80 (dd, $J_{H\text{-}H}$ = 5.4, 1.6 Hz, 2H, H$^{4'}$); 8.51 (d, $J_{H\text{-}H}$ = 8.4 Hz, 2H, H^2); 8.25 (dd, $J_{H\text{-}H}$ = 8.5, 5.2 Hz, 2H, H$^{3'}$); 8.11 (t, $J_{H\text{-}H}$ = 7.9, 2H, H^3); 7.96 (d, $J_{H\text{-}H}$ = 5.0 Hz, 2H, H^5); 7.16 (t, $J_{H\text{-}H}$ = 6.1 Hz, 2H, H^4); 6.11 (d, $^3J_{H\text{-}F}$ = 10.3 Hz, 2H, H^9). ^{13}C{^{1}H} NMR (100.6 MHz, acetone-d$_6$, δ): 184.4 (t, $^3J_{C\text{-}F}$ = 5.1 Hz, CHO); 164.4 (d, $^3J_{C\text{-}F}$ = 8.4 Hz, C^{10}); 164.2 (dd, $^1J_{C\text{-}F}$ = 235.4 Hz, $^3J_{C\text{-}F}$ = 6.3 Hz, C^8); 163.8 (d, $J_{C\text{-}F}$ = 6.6 Hz, C^{11} or C^{12}); 161.5 (dd, $^1J_{C\text{-}F}$ = 240.4 Hz, $^3J_{C\text{-}F}$ = 5.8 Hz, C^6); 154.2 (s, C$^{4'}$); 151.6 (s, C^5); 149.3 (s, C$^{9'}$); 148.1 (s, C^8); 141.4 (s, C^3); 140.7 (s, C$^{6'}$); 137.1 (s, C$^{2'}$); 132.1 (s, C$^{5'}$); 130.3 (s, C^{11} or C^{12}); 129.6 (s, C$^{3'}$); 126.0 (s, C^4); 125.5 (d, $^5J_{C\text{-}F}$ = 21.1 Hz, C^2); 116.2 (d, $^2J_{C\text{-}F}$ = 17.1 Hz, C^9); 111.2 (pst, $^2J_{C\text{-}F}$ = 12.0 Hz, C^7). ^{19}F {^{1}H} NMR (376.5 MHz, acetone-d$_6$, δ): −72.63 (d, $J_{F\text{-}P}$ = 707.7 Hz, 6F, PF$_6$); −114.23 (d, $^3J_{F\text{-}F}$ = 7.6 Hz, 2F, F^6); −117.43 (d, $^3J_{F\text{-}F}$ = 7.6 Hz, 2F, F^8).

3.3. Synthesis of [Ir(Si-Dfbzapy)$_2$(pyraphen)]PF$_6$ (2)

A solution of 0.10 g (0.09 mmol) of [Ir(dfbzapy)$_2$(pyraphen)]PF$_6$ (**1**) in 10 mL of THF was treated with 30.6 μL (0.20 mmol) of 3-aminopropyltriethoxysilane (APTES) in the presence of activated molecular sieves (used to collect the water produced during the reaction). The mixture was stirred for 1 hour at room temperature. The resulting yellow solution was filtered under an inert atmosphere and concentrated to ca. 3 mL. Due to the self-condensation tendency of complex **2** in the solid state, even under inert conditions, the compound cannot be isolated as a solid. Instead, the reaction mixture is directly used in the next step without further purification. The yield could not be determined. For the purposes of the subsequent reaction, a theoretical imination yield of 100% was assumed. ESI(+): m/z 1062 [{Ir(Si-dfbzapy)$_2$(pyraphen)} − 4Et − {2(-OEt)]$^+$ (100%). IR (cm^{-1}): ν(C-H aromatic) 2947 (m); ν(C=C, C=N) 1613 (s); ν(C-F) 1260; ν(Si–O–C) 1071 (vs); ν(Si–O) 1006 (vs); ν(P-F) 836 (vs); ν(Ir-N) 570 (s). ^{1}H NMR (300 MHz, acetone-d$_6$, δ): 9.83 (dd, $J_{H\text{-}H}$ = 8.3, 1.5 Hz, 2H, H$^{2'}$); 9.37 (s, 2H, H$^{8'}$); 8.81 (dd, $J_{H\text{-}H}$ = 5.2, 1.5 Hz, 2H, H$^{4'}$); 8.47 (m, 4H, N=CH y H^2); 8.27 (dd, $J_{H\text{-}H}$ = 8.4, 5.2 Hz, 2H, H$^{3'}$); 8.04 (t, $J_{H\text{-}H}$ = 7.8 Hz, 2H, H^3); 7.92 (d, $J_{H\text{-}H}$ = 6.1 Hz, 2H, H^5); 7.10 (pst, $J_{H\text{-}H}$ = 6.1 Hz, 2H, H^4); 5.99 (d, $^3J_{H\text{-}F}$ = 10.2, 2H, H^9); 3.81 (q, $J_{H\text{-}H}$ = 6.2 Hz, 12H, OCH_2CH$_3$); 3.63 (t, $J_{H\text{-}H}$ = 6.2 Hz, 2H, N–CH$_2$–); 3.16 (t, $J_{H\text{-}H}$ = 6.6 Hz, 2H, N–CH$_2$–); 1.83 (m, 2H, –CH$_2$–); 1.78 (m, 2H, –CH$_2$–); 1.19 (t, $J_{H\text{-}H}$ = 6.2 Hz, 18H, OCH$_2$CH_3); 0.66 (m, 4H, SiCH$_2$–). ^{19}F NMR (282 MHz, Acetone-d$_6$, δ): −72.60 (d, $J_{F\text{-}P}$ = 707.7 Hz, 6F, PF$_6$); −111.52 (d, $^3J_{F\text{-}F}$ = 9.1 Hz, 2F, F^6); −114.33 (d, $^3J_{F\text{-}F}$ = 9.2 Hz, 2F, F^8).

3.4. Synthesis of the Self-Condensed Material (SC-2)

As a continuation of the procedure for the preparation of [Ir(Si-dfppy)$_2$(pyraphen)]PF$_6$ (**2**) described above, the obtained solution of compound **2** in 3 mL of THF (approximately 0.09 mmol) was added dropwise to 13 mL of distilled water using an automated dropping system at a controlled rate of 0.02 mL/s to ensure thorough homogenization. After stirring at room temperature for 1 hour, 0.07 mL of a 0.05 M sodium fluoride solution was added, and the mixture was heated to 80 °C for 4 days. Upon completion of the reaction, the **SC-2** material precipitated as a yellow solid, which was isolated by centrifugation at

18,000 rpm for 30 min. The solid was subsequently washed with distilled water, ethanol, and acetonitrile, with centrifugation performed after each washing step. The washes with ethanol and acetonitrile effectively remove any unreacted complex **2**, resulting in the pure self-condensed material. (0.03 g, 30%). Anal. Calc. for $C_{48}H_{56}F_{10}IrN_8O_4Si_2P$: C, 45.60; H, 3.35; N, 8.86. Best analyses found: C, 39.20; H, 4.04; N, 5.22. (The self-condensed material is obtained in an aqueous medium. All the elemental analyses performed show similar values, maybe due to the presence of water and/or THF occluded in the material. In fact, the analyses were compatible with **SC-2**·8H_2O: C, 39.10; 3.58; N, 8.29). IR (cm^{-1}): ν(C-$H_{alifatic}$) 2963 (m); ν(C=C, C=N) 1600 (s); ν(C-F) 1259 (s); ν(Si–O–C) 1082 (vs); ν(Si–O) 1013 (vs); ν(P-F) 790 (vs); ν(Ir-N) 557 (s). MALDI (+): *m/z* 1065 [**SC-2** – {$(CH_2)_3Si(OEt)_3$} + 3H]$^+$ (1%); 861 [**SC-2** – {$(CH_2)_3Si(OEt)_3$} – (pyraphen) + Et + 2H] + (100%). MALDI (−): *m/z* 145 [PF_6]$^-$ (100%).

3.5. X-Ray Crystallographic Analysis of Complex **1**

Selected bond lengths and angles and X-ray crystallographic data for complex **1**·1.5$(CH_3)_2CO$ are summarized in SI (Tables S1 and S2). Orange crystals of complex **1** were obtained by slow diffusion at room temperature of *n*-hexane into a saturated solution of acetone. X-ray intensity data were collected with a Bruker D8 QUEST (PHOTON 100 CMOS) (Bruker, Karlsruhe, Germany) area-detector diffractometer, using graphite-monochromatic Mo-K_α radiation. The images were collected and processed using Bruker APEX4 and SAINT programs [38], carrying out the absorption correction at this point by semi-empirical methods using SADABS [38]. The structure was solved by intrinsic phasing using SHELXT [39], and refined by full-matrix least squares on F^2 with SHELXL [40], using the WINGX program suite [41]. All non-hydrogen atoms were assigned anisotropic displacement parameters. The inspection of the structure with PLATON [42] and SQUEEZE [43] revealed the presence of two voids of 310 e·Å^{-3} in the unit cell, each of them containing 99 electrons, which fits well for the presence of a total of six acetone molecules in the unit cell (one and half acetone molecule for each iridium(III) complex molecule—**1**·1.5$(CH_3)_2CO$). Supplementary crystallographic data for 1 were deposited at The Cambridge Crystallographic Data Centre (CCDC) under deposition number CCDC 2428185. These data can be obtained free of charge via www.ccdc.cam.ac.uk/data_request/cif (accessed on 4 April 2025).

3.6. Theoretical Calculations

Calculations for complex **2** (MeCN) were performed with the Gaussian 16 package [44], using the Becke three-parameter functional combined with the Lee–Yang–Parr correlation functional (B3LYP) [45] with the Becke–Johnson D3BJ correction [46]. No negative frequency was found in the vibrational frequency analysis of the final equilibrium geometries. The basis set used was the LanL2DZ effective core potential for Ir and 6-31G(d,p) for the ligand atoms [47]. DFT and TD-DFT calculations were performed using the polarized continuum model approach (PCM) [48], implemented in the Gaussian 16 software, in presence of the corresponding solvents. The predicted emission wavelengths were obtained from the energy difference between the triplet state at its optimized geometry and the singlet state in the triplet geometry. The results were visualized using GaussView 6. Overlap populations between molecular fragments were calculated using the GaussSum 3.0 software [49]. The emission energy was calculated as the difference of the optimized T_1 geometry for both states (adiabatic electronic transition).

3.7. Photocatalytic Methods for Dehalogenation of 2-Bromoacetophenone

Two different approaches were employed to study the photocatalytic reaction, based on the methodology previously described [26]:

In an NMR tube: A solution was prepared by dissolving 2-bromoacetophenone (15 mg, 0.075 mmol), triethanolamine (30 μL, 0.23 mmol), and 0.84 mg (0.75 μmol) of the photocatalyst **SC-2** in 0.8 ml of acetonitrile-d_3. The tube was degassed under nitrogen by bubbling for 5 min in the dark, followed by sonication for 1 min. The reaction was then irradiated with a 50 W blue LED (λ_{max} = 450 nm) for 9 min while stirring. The progress of the reaction was monitored by ^{1}H-NMR, with 1,3,5-trimethoxybenzene (0.06 mmol) as the internal standard, to determine the yield.

In a round bottom flask: Under a nitrogen atmosphere, 2-bromoacetophenone (0.65 g, 3.27 mmol), **SC-2** (36.51 mg, 0.03 mmol), and 15 mL of anhydrous acetonitrile (previously degassed by nitrogen bubbling for 20 min) were added to a 50 mL round-bottom flask. After sonicating the mixture for 1 min, triethanolamine (1.3 mL, 9.8 mmol) and 1,3,5-trimethoxybenzene (300 mg, 1.78 mmol) were added. The reaction mixture was irradiated with a 50 W blue LED (λ_{max} = 450 nm) for 90 min while being stirred, and the progress of the reaction was also monitored by ^{1}H-NMR.

For the recovery and reusing of the catalyst **SC-2**, the mixture was centrifuged at 18,000 rpm for 30 min and the resulting solid was washed with distilled water and sonicated for 10 min, followed by centrifugation at 18,000 rpm for another 30 min. This washing and centrifugation process was repeated using absolute ethanol and then acetonitrile. The solid was subsequently suspended in dichloromethane, transferred back to the reaction flask, and dried under vacuum at 100 °C to remove any residual moisture. The same procedure was applied to subsequent reactions. Yields were determined using ^{1}H-NMR analysis with 1,3,5-trimethoxybenzene serving as the internal standard.

4. Conclusions

In this study, we describe the synthesis and characterisation of a new photocatalytic material based on the Ir(III) complex [Ir(Si-dfbzapy)$_2$(pyraphen)]PF_6 (**2**), functionalized with alkoxysilane groups in the cyclometalated ligands. This complex undergoes self-condensation, leading to the formation of the organometallo–ionosilica **SC-2**. Structural and spectroscopic analyses confirm the preservation of the structure of the cationic Ir(III) complex after the heterogenization process, as well as the homogeneous distribution of the PF_6^- counteranion throughout the material, highlighting its ionic nature.

From a photophysical point of view, while **SC-2** exhibits a weak red emission in the solid state, its absorption and emission spectra in acetonitrile suspension are practically identical to those of its precursor **2** in solution in the same solvent, showing a blue greenish phosphorescence and a significant absorption between 400 and 500 nanometers. In addition, the electrochemical properties calculated for **SC-2** and complex **2**, with the latter acting as a model of the material in solution, indicate its feasibility for photoinduced electron transfer processes. Thus, the **SC-2** material has shown excellent photocatalytic activity in heterogeneous conditions in the reductive dehalogenation reaction of 2-bromoacetophenone under blue light illumination (450 nm). In terms of reusability, **SC-2** retains its photocatalytic activity over seven catalytic cycles, with a decrease in efficiency under 30% and likely due to material loss during the washing steps. Nevertheless, post-reaction analysis of the material suggests the possibility of some photodegradation of surface-bound molecules during irradiation. Despite this, the material exhibits remarkable photostability, allowing it to be used in multiple catalytic cycles.

Overall, these findings highlight the potential of **SC-2** as an efficient, sustainable, and reusable heterogeneous photocatalyst for applications in organic synthesis and solar-driven redox transformations.

Supplementary Materials: The following supporting information can be downloaded at: https://www.mdpi.com/article/10.3390/molecules30081680/s1. Supplementary Materials includes characterization data for complexes **1** and **2**, and material **SC-2**, including the photophysical and electronic properties, the theoretical calculations performed on complex **2**, and the photostability studies of **SC-2**.

Author Contributions: J.H.: Formal Analysis, Investigation, Methodology, Theoretical Calculations, Validation, Writing—original draft. M.M.-A.: Formal analysis, Investigation, Methodology, Validation. J.G.-B.: Investigation, Software. M.A.R.: Conceptualization, Funding acquisition, Project administration, Supervision, Theoretical Calculations, Writing—review and editing. J.R.B.: Conceptualization, Funding acquisition, Project administration, Supervision, Writing—review and editing. All authors have read and agreed to the published version of the manuscript.

Funding: This research was funded by the Spanish MCIN/AIE/10.13039/501100011033, the "ERDF A way of making Europe", the "European Union" (project PID2023-149547NB-I00), and Gobierno de la Rioja (Project FORTALECE 2021/01).

Data Availability Statement: Data will be made available on request.

Acknowledgments: This work was supported by the Spanish MCIN/AIE/10.13039/501100011033, the "ERDF A way of making Europe", the "European Union" (project PID2023-149547NB-I00), and Gobierno de la Rioja (Project FORTALECE 2021/01). We would like to thank Elena Serrano and Javier García-Martínez from the University of Alicante for their generous help with the characterization of the **SC-2** material.

Conflicts of Interest: The authors declare no conflicts of interest.

References

1. Kou, J.; Lu, C.; Wang, J.; Chen, Y.; Xu, Z.; Varma, R.S. Selectivity Enhancement in Heterogeneous Photocatalytic Transformations. *Chem. Rev.* **2017**, *117*, 1445–1514. [CrossRef]
2. Ham, R.; Nielsen, C.J.; Pullen, S.; Reek, J.N.H. Supramolecular Coordination Cages for Artificial Photosynthesis and Synthetic Photocatalysis. *Chem. Rev.* **2023**, *123*, 5225–5261. [CrossRef] [PubMed]
3. Chandra, P.; Choudhary, N.; Mobin, S.M. The game between molecular photoredox catalysis and hydrogen: The golden age of hydrogen budge. *Mol. Catal.* **2023**, *537*, 112921. [CrossRef]
4. Millward, F.; Zysman-Colman, E. Alchemy reimagined: Photocatalysis using anthropogenic waste materials. *Trends Chem.* **2023**, *5*, 584–587. [CrossRef]
5. Ciamician, G.; Silber, P. Chemische Lichtwirkungen. *Ber. Dtsch. Chem. Ges.* **1908**, *41*, 1928–1935. [CrossRef]
6. Heindel, N.D.; Pfau, M.A. A profitable partnership: Giacomo Ciamician and Paul Silber. *J. Chem. Educ.* **1965**, *42*, 383. [CrossRef]
7. Yakushev, A.A.; Abel, A.S.; Averin, A.D.; Beletskaya, I.P.; Cheprakov, A.V.; Ziankou, I.S.; Bonneviot, L.; Bessmertnykh-Lemeune, A. Visible-light photocatalysis promoted by solid- and liquid-phase immobilized transition metal complexes in organic synthesis. *Coord. Chem. Rev.* **2022**, *458*, 214331. [CrossRef]
8. Marzo, L.; Pagire, S.K.; Reiser, O.; König, B. Visible-Light Photocatalysis: Does It Make a Difference in Organic Synthesis? *Angew. Chem. Int. Ed.* **2018**, *57*, 10034–10072. [CrossRef]
9. König, B. Photocatalysis in Organic Synthesis—Past, Present, and Future. *Eur. J. Org. Chem.* **2017**, *2017*, 1979–1981. [CrossRef]
10. Lindroth, R.; Materna, K.L.; Hammarström, L.; Wallentin, C.-J. Sustainable Ir-Photoredox Catalysis by Means of Heterogenization. *ACS Org. Inorg. Au* **2022**, *2*, 427–432. [CrossRef] [PubMed]
11. Chan, A.Y.; Perry, I.B.; Bissonnette, N.B.; Buksh, B.F.; Edwards, G.A.; Frye, L.I.; Garry, O.L.; Lavagnino, M.N.; Li, B.X.; Liang, Y.; et al. Metallaphotoredox: The Merger of Photoredox and Transition Metal Catalysis. *Chem. Rev.* **2022**, *122*, 1485–1542. [CrossRef] [PubMed]
12. Prier, C.K.; Rankic, D.A.; MacMillan, D.W.C. Visible Light Photoredox Catalysis with Transition Metal Complexes: Applications in Organic Synthesis. *Chem. Rev.* **2013**, *113*, 5322–5363. [CrossRef] [PubMed]
13. Bevernaegie, R.; Wehlin, S.A.M.; Elias, B.; Troian-Gautier, L. A Roadmap Towards Visible Light Mediated Electron Transfer Chemistry with Iridium(III) Complexes. *ChemPhotoChem* **2021**, *5*, 217–234. [CrossRef]
14. Bell, J.D.; Murphy, J.A. Recent advances in visible light-activated radical coupling reactions triggered by (i) ruthenium, (ii) iridium and (iii) organic photoredox agents. *Chem. Soc. Rev.* **2021**, *50*, 9540–9685. [CrossRef]
15. Glaser, F.; Wenger, O.S. Recent progress in the development of transition-metal based photoredox catalysts. *Coord. Chem. Rev.* **2020**, *405*, 213129. [CrossRef]

16. Singh, A.; Teegardin, K.; Kelly, M.; Prasad, K.S.; Krishnan, S.; Weaver, J.D. Facile synthesis and complete characterization of homoleptic and heteroleptic cyclometalated Iridium(III) complexes for photocatalysis. *J. Organomet. Chem.* **2015**, *776*, 51–59. [CrossRef]
17. Tritton, D.N.; Tang, F.-K.; Bodedla, G.B.; Lee, F.-W.; Kwan, C.-S.; Leung, K.C.-F.; Zhu, X.; Wong, W.-Y. Development and advancement of iridium(III)-based complexes for photocatalytic hydrogen evolution. *Coord. Chem. Rev.* **2022**, *459*, 214390. [CrossRef]
18. Sakai, K.; Ozawa, H. Homogeneous catalysis of platinum(II) complexes in photochemical hydrogen production from water. *Coord. Chem. Rev.* **2007**, *251*, 2753–2766. [CrossRef]
19. Shi, J.; Su, Z.; Li, X.; Feng, J.; Men, C. Impacts of host–guest assembly on the photophysical and photocatalytic properties of heterogenized molecular photosensitizer and catalysts. *J. Mater. Chem. A* **2023**, *11*, 6646–6658. [CrossRef]
20. Kunzmann, A.; Valero, S.; Sepúlveda, Á.E.; Rico-Santacruz, M.; Lalinde, E.; Berenguer, J.R.; García-Martínez, J.; Guldi, D.M.; Serrano, E.; Costa, R.D. Hybrid Dye-Titania Nanoparticles for Superior Low-Temperature Dye-Sensitized Solar Cells. *Adv. Energy Mater.* **2018**, *8*, 1702583. [CrossRef]
21. Wang, Z.; Li, C.; Domen, K. Recent developments in heterogeneous photocatalysts for solar-driven overall water splitting. *Chem. Soc. Rev.* **2019**, *48*, 2109–2125. [CrossRef] [PubMed]
22. Luo, T.; Gilmanova, L.; Kaskel, S. Advances of MOFs and COFs for photocatalytic CO_2 reduction, H_2 evolution and organic redox transformations. *Coord. Chem. Rev.* **2023**, *490*, 215210. [CrossRef]
23. Li, Y.-L.; Li, A.-J.; Huang, S.-L.; Vittal, J.J.; Yang, G.-Y. Polypyridyl Ru(ii) or cyclometalated Ir(iii) functionalized architectures for photocatalysis. *Chem. Soc. Rev.* **2023**, *52*, 4725–4754. [CrossRef]
24. Hao, Y.; Lu, Y.-L.; Jiao, Z.; Su, C.-Y. Photocatalysis Meets Confinement: An Emerging Opportunity for Photoinduced Organic Transformations. *Angew. Chem. Int. Ed.* **2024**, *63*, e202317808. [CrossRef] [PubMed]
25. Martínez-Aguirre, M.; Serrano, E.; Ezquerro, C.; Lalinde, E.; Berenguer, J.R.; García-Martínez, J.; Rodríguez, M.A. Hybrid organometallo-silica catalysts for sustainable visible-light promoted olefin isomerization. *Catal. Today* **2023**, *422*, 114213. [CrossRef]
26. Martínez-Aguirre, M.; Herce, J.; Serrano, E.; García-Martínez, J.; Rodríguez, M.A.; Berenguer, J.R. Self-condensed organometallo Ir(III) ionosilica for sustainable visible-light promoted electron-transfer photocatalysis. *J. Catal.* **2025**, *443*, 115946. [CrossRef]
27. Tran Duy, T.; Duarte Rodrigues, A.; Vo-Thanh, G.; Hesemann, P. Dialkyl imidazolium acetate ionosilica as efficient and recyclable organocatalyst for cyanosilylation reactions of ketones. *Green Energy Environ.* **2020**, *5*, 130–137. [CrossRef]
28. Wu, H.; Hesemann, P.; Trens, P.; Silly, G.; Salles, F.; Zajac, J. Ionosilica-based anion exchangers for low-temperature thermochemical storage of energy under mild conditions of adsorbent regeneration and saturation. *Chem. Eng. J.* **2020**, *398*, 125634. [CrossRef]
29. Hesemann, P. Applications of ionosilicas in heterogeneous catalysis: Opportunities for the elaboration of new functional catalytic phases. *Curr. Opin. Green Sustain. Chem.* **2018**, *10*, 21–26. [CrossRef]
30. Materna, K.L.; Hammarström, L. Photoredox Catalysis Using Heterogenized Iridium Complexes. *Chem. Eur. J.* **2021**, *27*, 16966–16977. [CrossRef]
31. Zhang, W.; Wang, D.; Xie, Q.; Xu, C.; Kuang, G.; Tang, J.; Pan, C.; Yu, G. Enhanced Reducibility via Altering Exciton Binding Energy of Conjugated Microporous Polymers for Photocatalytic Reduction. *Macromolecules* **2023**, *56*, 4022–4029. [CrossRef]
32. Devika, S.; Palanichamy, M.; Murugesan, V. Selective oxidation of ethylbenzene over CeAlPO-5. *Appl. Catal. A* **2011**, *407*, 76–84. [CrossRef]
33. Okamura, N.; Nakamura, T.; Yagi, S.; Maeda, T.; Nakazumi, H.; Fujiwara, H.; Koseki, S. Novel bis- and tris-cyclometalated iridium (iii) complexes bearing a benzoyl group on each fluorinated 2-phenylpyridinate ligand aimed at development of blue phosphorescent materials for OLED. *RSC Adv.* **2016**, *6*, 51435–51445. [CrossRef]
34. Millán, G.; Nieddu, M.; López, I.P.; Ezquerro, C.; Berenguer, J.R.; Larráyoz, I.M.; Pichel, J.G.; Lalinde, E. A new family of luminescent iridium complexes: Synthesis, optical, and cytotoxic studies. *Dalton Trans.* **2023**, *52*, 6360–6374. [CrossRef]
35. Ezquerro, C.; Sepulveda, A.E.; Grau-Atienza, A.; Serrano, E.; Lalinde, E.; Berenguer, J.R.; Garcia-Martinez, J. Organometallic phosphors as building blocks in sol-gel chemistry: Luminescent organometallo-silica materials. *J. Mater. Chem. C* **2017**, *5*, 9721–9732. [CrossRef]
36. Ezquerro, C.; López, I.; Serrano, E.; Alfaro-Arnedo, E.; Lalinde Peña, E.; Larráyoz, I.; Pichel, J.G.; García-Martinez, J.; Berenguer, J.R. Highly Emissive Hybrid Mesoporous Organometallo-Silica Nanoparticles for Bioimaging. *Mater. Adv.* **2022**, *3*, 3582–3592. [CrossRef]
37. Bryden, M.A.; Millward, F.; Lee, O.S.; Cork, L.; Gather, M.C.; Steffen, A.; Zysman-Colman, E. Lessons learnt in photocatalysis—The influence of solvent polarity and the photostability of the photocatalyst. *Chem. Sci.* **2024**, *15*, 3741–3757. [CrossRef]
38. Bruker AXS Inc. *APEX4, SAINT, and SADABS*; Bruker AXS Inc.: Fitchburg, WI, USA, 2021.
39. Sheldrick, G.M. SHELXT—Integrated space-group and crystal-structure determination. *Acta Crystallogr. Sect. A Found. Adv.* **2015**, *71*, 3–8. [CrossRef]
40. Sheldrick, G.M. Crystal structure refinement with SHELXL. *Acta Crystallogr. Sect. C Struct. Chem.* **2015**, *71*, 3–8. [CrossRef]
41. Farrugia, L.J. WinGX suite for small-molecule single-crystal crystallography. *J. Appl. Crystallogr.* **1999**, *32*, 837–838. [CrossRef]

42. Spek, A.L. Single-crystal structure validation with the program PLATON. *J. Appl. Crystallogr.* **2003**, *36*, 7–13. [CrossRef]
43. Spek, A.L. PLATON SQUEEZE: A tool for the calculation of the disordered solvent contribution to the calculated structure factors. *Acta Crystallogr. Sect. C Struct. Chem.* **2015**, *71*, 9–18. [CrossRef]
44. Frisch, M.J.; Trucks, G.W.; Schlegel, H.B.; Scuseria, G.E.; Robb, M.A.; Cheeseman, J.R.; Scalmani, G.; Barone, V.; Petersson, G.A.; Nakatsuji, H. *Gaussian 16. 3*; Gaussian, Inc.: Wallingford, CT, USA, 2016.
45. Becke, A.D. Density-functional thermochemistry. III. The role of exact exchange. *J. Chem. Phys.* **1993**, *98*, 5648–5652. [CrossRef]
46. Grimme, S.; Antony, J.; Ehrlich, S.; Krieg, H. A consistent and accurate ab initio parametrization of density functional dispersion correction (DFT-D) for the 94 elements H-Pu. *J. Chem. Phys.* **2010**, *132*, 154104. [CrossRef] [PubMed]
47. Wadt, W.R.; Hay, P.J. Ab initio effective core potentials for molecular calculations. Potentials for main group elements Na to Bi. *J. Chem. Phys.* **1985**, *82*, 284–298. [CrossRef]
48. Barone, V.; Cossi, M. Quantum Calculation of Molecular Energies and Energy Gradients in Solution by a Conductor Solvent Model. *J. Phys. Chem. A* **1998**, *102*, 1995–2001. [CrossRef]
49. O'Boyle, N.M.; Tenderholt, A.L.; Langner, K.M. cclib: A library for package-independent computational chemistry algorithms. *J. Comput. Chem.* **2008**, *29*, 839–845. [CrossRef]

Disclaimer/Publisher's Note: The statements, opinions and data contained in all publications are solely those of the individual author(s) and contributor(s) and not of MDPI and/or the editor(s). MDPI and/or the editor(s) disclaim responsibility for any injury to people or property resulting from any ideas, methods, instructions or products referred to in the content.

molecules

MDPI

Article

Organic–Inorganic Hybrid Ladder-like Polysilsesquioxanes as Compatibilized Nanofiller for Nanocomposite Materials

Dominique Mouysset [1], Marion Rollet [1], Emily Bloch [2], Stéphane Gastaldi [1], Eric Besson [1,*] and Trang N. T. Phan [1,*]

1 Aix Marseille Univ, CNRS, Chemistry Department, Institute of Radical Chemistry (ICR), 13397 Marseille, France; dominique.mouysset@univ-amu.fr (D.M.); marion.rollet@univ-amu.fr (M.R.); stephane.gastaldi@univ-amu.fr (S.G.)

2 Aix Marseille Univ, CNRS, Chemistry Department, Laboratory of Divided Materials, Interfaces, Reactivity, Electrochemistry (MADIREL), 13397 Marseille, France; emily.bloch@univ-amu.fr

* Correspondence: eric.besson@univ-amu.fr (E.B.); trang.phan@univ-amu.fr (T.N.T.P.)

Abstract: Nanocomposite materials composed of an organic matrix and an inorganic nanofiller have been the subject of intense research in recent years. Indeed, the synergy between these two phases confers improved properties thanks to an increased surface–volume ratio, which reinforces the interactions between the particles and the polymer matrix. These interactions depend on many factors such as the shape, size and dispersion of the nanoobjects. Polysilsesquioxanes (PSQs) are a silicon polymer family that offers different sizes, shapes and structures and possesses ceramics properties (i.e., high thermal and/or oxidative resistance and high chain rigidity), thanks to the siloxane backbone. In this article, we propose to incorporate polymer-grafted ladder polysilsesquioxanes (LPSQs) as nanofillers in thermoplastic matrices. Chloride-functionalized LPSQs were synthesized from two different precursors and thoroughly characterized by ^{1}H, ^{13}C and ^{29}Si NMR, as well as by SEC and WAXS. The well-defined LPSQ was then converted into an azide analog. The resulting hybrid material was functionalized with poly(ethylene glycol) (PEG) chains and incorporated into poly(ethylene oxide) or poly(methyl methacrylate) matrices. We found that the viscoelastic properties of the nanocomposite materials were impacted by plasticizing or the reinforcement effect depending on the grafted PEG chain length.

Keywords: ladder polysilsesquioxane; PEG-grafted LPSQ; nanocomposites

1. Introduction

With a global production of more than 300 million tons, synthetic polymers occupy an increasingly important place in many objects and materials in everyday life. They are found in all sectors as commodity materials (packaging, transport, sports and leisure, etc.) or as specialty polymers where they perform more "noble" functions in fields such as electronics, optics, biomedical and cosmetics. In recent decades, new hybrid materials have been developed. These materials aim to exploit the complementary properties of their components or to create synergy between them; we then speak of composites. In other words, a composite material is a material for which the combination of basic components leads to a material having properties that none of the components taken separately has. They can be defined as a material resulting from the association of a continuous phase, the matrix, and a dispersed phase, strengthening. Depending on the composition of composite materials, they can have many properties and therefore many advantages. Composite materials allow almost infinite freedom of assembly, which makes it possible to create on demand materials with the desired properties depending on the product manufactured. Today, there are a large number of composite materials that are generally classified into three families depending on the nature of the matrix, which can be ceramic, metallic or organic. Organic-based composites are the most important class of these materials thanks to their advantages such as ease of implementation, low cost, resistance to corrosion and their major advantage compared to metallic compounds, their low weight. These materials are prepared by adding fillers, generally inorganic for an organic matrix, in the form of a particle or fiber. However, the efficiency of the performances expected by the addition of these charges is linked to the quantity of surface area of the interfaces, generally requiring a significant concentration of charge. The properties of the polymer matrix in terms of processability are strongly affected. This problem found a solution in the early 1990s, with the pioneering work of Toyota researchers by the incorporation of a small amount (a few percent by mass) of lamellar clay in a polyamide [1]. These new composite materials, named nanocomposites because at least one of the phases has at least one dimension less than 100 nanometers, have seen their mechanical properties greatly improved. Indeed, the very small dimensions of these nanofillers allow increasing the surface-to-volume ratio, reinforcing the degree of particle–matrix interactions.

Thus, the nanocomposites' properties and the filler/polymer interactions depend on several factors such as the shape and size, the quantity and dispersion quality, the orientation and distribution of the nanoobjects. Of course, all are connected together and to ensure a good transfer of the properties of the nanomaterial to the polymeric matrix, it is necessary to control its shape (for a high aspect ratio) [2] and the interface [3]. For the latter, generally, a modification of the nanofiller by macromolecules, identical or compatible, to those of the matrix is carried out in order to increase their affinities and the quality of dispersion.

Among the different fillers, silsesquioxane (SSQs) or polysilsesquioxane (PSQ) have attracted particular attention because they can be functionalized with a large diversity of organic groups and are inorganic materials with a high compatibility with organic materials. PSQs are of general formula [R-$SiO_{1,5}$] and can be prepared in many shapes, such as random, ladder (LPSQs), lamellar, cage (POSS) or partial cage structures [4–6].

Considered as the smallest hybrid particles of silica, POSS was certainly the most SSQs used for the preparation of nanocomposite materials [7]. They can be incorporated in the polymer matrix as non-functional (reinforcement agent), monofunctional (polymer end-capping agent) or multifunctional (crosslinking agent) POSS. Similarly, we have recently prepared a new alkoxyamine-functionalized lamellar PSQ hybrid material, exfoliated it and used it both as a filler and polymerization initiator in a grafting from strategy. The presence of polysilsesquioxane in the composites had an impact on their viscoelastic properties, demonstrating the reinforcement of the polymeric matrix by the lamellae [8]. LPSQ is another SSQ with a tailorable and controlled structure described in these last few years because it addresses some POSS limitations (solubility, low molecular weight...) [9]. In addition to the thermal/chemical stability of the siloxane backbone, the double-stranded structure contributes to the robustness of the LPSQ backbone. The work of Lee is a good example [10]. They systematically studied the effect of the SSQ structure on the mechanical properties of SSQ-reinforced polymer film. The results suggest that the highly ordered and one-dimensional LPSQ, with a higher aspect ratio, leads to a better enhancement in the mechanical properties than random and POSS SQs. In these composites, the SSQ structures were covalently linked to the polymer matrix via the photopolymerization of methacrylate groups grafted on SSQ structures and an acrylate resin used as the polymer matrix. Another approach to improve the compatibility of LPSQ and polyimide was studied by Kim et al., which is based on the hydrogen-bonding interactions between PEG-grafted LPSQ and the carboxylic acid groups of polyimide [11]. Nevertheless, the LPSQ used was also functionalized with methacrylate groups and copolymerized by UV-curing with PEG-dimethacrylate presence in the polyimide composite formulation. As mentioned above, one main strategy to improve the dispersion of fillers such as silica nanoparticles within polymeric matrices is to functionalize their surfaces with polymers identical or compatible to those of the matrix [12]. This approach is easy to implement and an efficient way to enhance the compatibility between the nanofiller and organic polymer matrix, but it has rarely been exploited with LPSQs in the literature.

In the aim to enlarge the development of LPSQ as a nanofiller and simplify its composite elaboration while ensuring their good dispersion within the commercial polymer matrix, we studied, in the present work, the synthesis of well-defined poly(ethylene glycol) (PEG)-grafted LPSQs and their use as filler in two different commercial polymer matrices, namely poly(ethylene oxide) (PEO) and poly(methyl methacrylate) (PMMA). PEO is one of the most studied polymer electrolytes for lithium battery application [13]. This is because PEO presents several advantages such as good flexibility, ease of processing, excellent interfacial stability with lithium electrode and the formation of complexes with lithium salts. However, the commercialization of PEO-based electrolytes in lithium battery is difficult because their room-temperature (RT) ionic conductivity is too low due to the high degree of crystallization of PEO at RT that reduces the chain mobility and, hence, lithium mobility as well as ionic conductivity. We expect that the incorporation of PEG-grafted LPSQ allows for reducing the PEO crystallinity while maintaining or improving its mechanical properties. Therefore, it could be an interesting approach to increase the ionic conductivity of PEO-based electrolytes. On the other hand, blending polymers is an efficient and economic method to develop new polymeric materials. PEO/PMMA is a partially miscible crystalline/amorphous blend, which has attracted much interest [14–19]. Currently, there are no papers investigating the blends of PMMA and grafted PEO and their viscoelastic properties. So, it could be interesting to study the behavior of PEG-grafted LPSQ and PMMA blends. For these reasons, we first synthesized chloride-functionalized LPSQ from two different precursors. The well-defined LPSQ was chosen for conversion into an azide analog before grafting with PEG chains by the "grafting to" approach. Secondly, the resulting PEG-grafted LPSQ hybrid materials were blended at different contents with PEO or PMMA. It is expected that the presence of grafted PEG on LPSQ will increase their affinities/compatibilities with the polymer matrix, and hence the quality of dispersion, in order to ensure a good transfer of the properties of the hybrid material to the polymeric matrix.

2. Results and Discussion

2.1. Synthesis of Ladder Hybrid Materials

The synthesis of functionalized ladder-like polysilsesquioxanes (LPSQs) was implemented from 3-chloropropyl trialkoxysilane **1a** and **1b** in order to assess the impact of their alkoxy groups, i.e., ethoxy or methoxy, on the features of the LPSQs (Scheme 1) [20]. **1a** or **1b** was stirred at room temperature in a water/THF mixture in the presence of a catalytic amount of K_2CO_3 (Scheme 1) [21]. In the case of **1a**, MeOH (0.5 equiv.) was added for triggering silanol formation. After only 18 h, NMR monitoring showed the completion of the polymerization of **1b**. This result is in agreement with a fast hydrolysis of trimethoxysilane moiety. The presence of insoluble PSQ in the crude solution resulted from random polymerization; its filtration yielded LPSQ **2b** in 26% yield. As expected, in the case of **1a**, the reaction was much slower, and 9 days were needed to reach 96% of the polymerization of **1a** (NMR monitoring). Under these slow polymerization conditions, no insoluble material, i.e., random polymerization, was detected. After work-up, **2a** was obtained quantitatively.

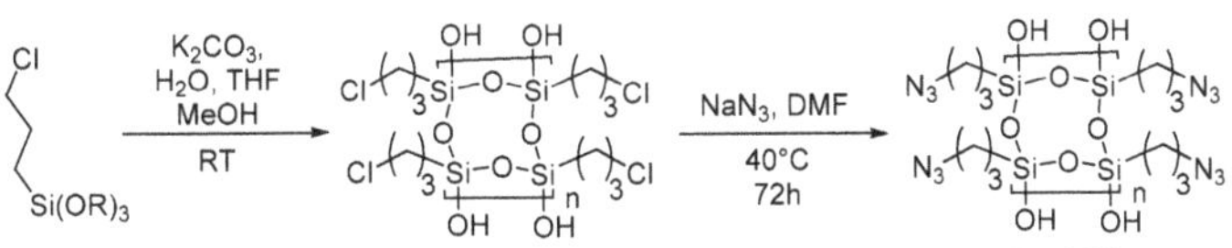

1a R = Et (9 days).
1b R = Me (18 h).
2a 100%
2b 26%
3 86%

Scheme 1. Synthesis of functionalized ladder materials.

The solubility of **2a** and **2b** in several organic solvents allows performing the liquid NMR analysis in $CDCl_3$. The ^1H and ^{13}C spectra confirmed the integrity of the organic moieties under the sol–gel process conditions, as shown by the presence of characteristic signals at 3.52 (CH_2Cl) and 0.79 ppm (CH_2Si) recorded for the two LPSQs.

Solid-state ^{29}Si NMR (SS NMR) is one of the key analyses for asserting the ladder structure of the PSQ. Indeed, highlighting the difference between the ratio of T^1 (RS*i*(OH)$_2$OSi), T^2 (RS*i*(OH)(OSi)$_2$) and T^3 (RS*i*(OSi)$_3$) siloxane structures allows revealing the level of siloxane bond formation [9]. The analysis of the ^{29}Si NMR spectra for **2a** and **2b** showed the presence of T^1, T^2 and T^3 signals of very weak, weak and strong intensity, respectively (Figures S1 and S2, Supplementary Materials). PSQ **2a** and **2b** exhibited a high degree of condensation (DOC) of 98.8 and 97.5%, consistent with an LPSQ structure [9].

The second key analysis is the Fourier-transform infrared (FTIR) spectroscopy. Indeed, a characteristic double peak at 1100 and 1026 cm^{-1} is observed and attributed to the asymmetric and symmetric stretching vibration of Si–O–Si, which is directly linked to the LPSQ structure (Figure S3, Supplementary Materials) [22]. Furthermore, the absence of peaks corresponding to the silanol group (Si-OH) at 3500 cm^{-1} and 960 cm^{-1} was in accord with ^{29}Si SS NMR, which exhibited a very weak T^1 signal and weak peak for T^2 in the case of **2a**, confirming the high level of condensation of the silanol to form siloxane bridges.

Powder wide-angle X-ray scattering (WAXS) analysis was carried out on **2a** and gives a good insight into the structure of the polysilsesquioxane. The WAXS pattern (Figure 1) exhibits two maxima at 5.4 and 14.8 nm^{-1}. The first maximum corresponds to distance d1 = 1.16 nm, attributed to the chain-to-chain distance, and the second d2 = 0.42 nm is related to the average thickness of the double-stranded siloxane backbone. Whereas random polysilsequioxanes generally showed a unique broad peak due to low regularity and POSS displayed sharp peaks, these results are in accordance with ladder-like polysilsesquioxane [23].

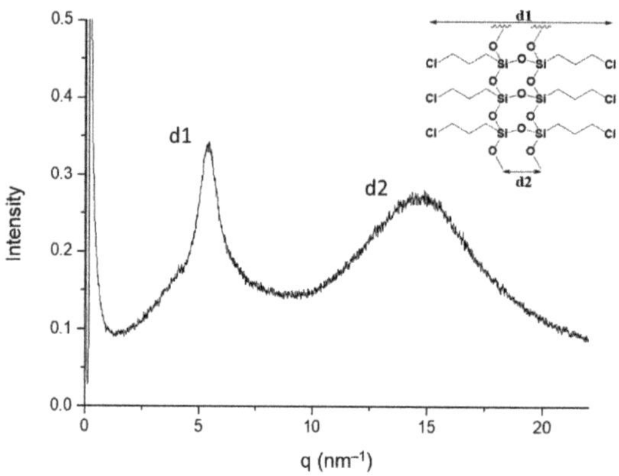

Figure 1. WAXS measurement of LPSQ **2a**.

SEC analysis was used to determine the M_w. The LPSQ **2b**, prepared from trimethoxysilane **1b**, had a slightly lower M_w of 8200 g mol^{-1} than that measured for **2a** of 10,500 g mol^{-1} and a comparable dispersity (Đ) of 2.4 and 2.5, respectively.

These analyses clearly pointed to a well-defined structure for **2a**. The slower polycondensation with the triethoxysilane **1a** provided an LPSQ with a low level of defects, a higher molecular weight and a higher yield. Therefore, LPSQ **2a** was used for the remainder of this study.

The second step in the preparation of the hybrid material was the introduction of an azide function via a nucleophilic substitution, so that in the next step, a poly(ethylene glycol) chain could be introduced via a click reaction. A solution of **2a** in DMF was stirred at 40 °C for 72 h in the presence of 1.5 equivalents of NaN_3. N_3-LPSQ **3** was obtained with a yield of 86%. ^{1}H and ^{13}C NMR confirmed a total conversion as no CH_2Cl signals were detected and new signals corresponding to CH_2N_3 appeared at 3.28 and 53.5 ppm, respectively.

The FTIR analysis also established the presence of an azide function (2090 cm^{-1}) and the stability of the ladder structure under substitution conditions, thanks to the two peaks at 1096 and 1040 cm^{-1}. It is noteworthy that the ^{29}Si NMR spectrum now displayed a complete disappearance of the T^1 signal, as shown by the DOC of 99.7%, calculated from the integration of T^n signals. It is likely that the aforementioned substitution conditions also favored the formation of the missing siloxane bonds in the ladder structure. At this point, silanol polycondensation can be considered complete.

The last step in the synthesis of the ladder-based hybrid material was the grafting of a polyethylene glycol (PEG) chain through a click reaction between a triple bond and an azide function. This reaction was performed with two propargylated PEG with molecular weights of 1000 g mol^{-1} (**4**) and 5000 g mol^{-1} (**5**) (Scheme 2). Two different experimental conditions were implemented to form the 1,2,3-triazole ring. In the first approach, the copper sulfate/sodium ascorbate couple was used to catalyze the click reaction [24,25], while the second method involved a thermal activation to enable the [2+3] cycloaddition [24].

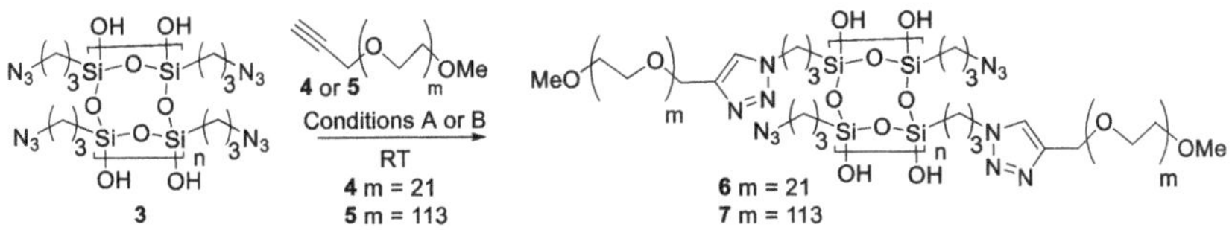

Conditions A: **3** (1eq), **4** (1 eq.), $CuSO_4$ (2 eq.)/ sodium ascorbate (2 eq.), H_2O/THF, RT, 10 days
Conditions B: **3** (1eq.), **5** (0.1 eq.), 130°C, 48 h

Scheme 2. Synthesis of poly(ethylene glycol)-grafted ladder materials.

In both cases, the reaction advancement was monitored by ^1H NMR with the disappearance of the Csp-H signal at 2.4 ppm from the triple bond of **4** or **5**. It is worth pointing out that the ^{13}C NMR analysis of **6** (PEG1-LPSQ) and **7** (PEG5-LPSQ) showed that the copper-catalyzed cyclization led only to 1,4-disubstituted 1,2,3-triazole (C quaternary = 143.5 ppm, Csp^2-H = 123.1 ppm) whereas the thermal conditions provided a 1:1 mixture of 1,4- and 1,5-disubstituted rings (C quaternary and Csp^2H = 134.5 ppm). TGA analysis was used to determine the grafting yield of the PEG chains, which was 56 mol % for **6** and 4 mol % for **7** as well as the weight fraction of PEG in these hybrid materials, which was 80 wt% and 61 wt%, respectively. The calculations are detailed in Figures S4 and S5 (Supplementary Materials). Although the grafting did not involve all the azide functions, it can be noted that, in the case of PEG1-LPSQ (**6**), the characteristic FTIR absorption band attributed to the azide group at 2090 cm^{-1} was weak compared to that which should be expected while that of PEG5-LPSQ (**7**) is still observable (Figure S6a,b, Supplementary Materials). Concerning the ^{13}C NMR analysis, the same observation was made for PEG1-LPSQ (**6**) and PEG5-LPSQ (**7**); in other words, the signal corresponding to CH_2-N_3 around 53 ppm is weak or even undetectable for the PEG1-LPSQ (**6**) compound and well present in the spectra of PEG5-LPSQ (**7**) (Figure S7a,b, Supplementary Materials). Differential scanning calorimetry (DSC) was used to analyze the thermal behavior of grafted PEG and look at how the presence of the LPSQ, covalently linked to PEG chains, affects the melting temperature (T_m) and the crystallinity degree (X_c) of PEG. The pristine PEG of 1000 g mol^{-1} and 5000 g mol^{-1} exhibit a T_m of 38.7 and 64.3 °C, respectively. The PEG-grafted LPSQ showed a T_m of 36.5 and 57.3 °C for **6** and **7**, respectively, which are slightly lower than those of the corresponding pristine PEG. From the heat fusion of the DSC thermogram (Figures S8 and S9, Supplementary Materials), the crystallinity degree of PEG was calculated using Equation (1) and reported in Table 1. We observe a significant reduction in X_c of the grafted PEG compared to the pristine PEG of the same molar mass and this trend is more pronounced with the PEG of low molecular weight. Since the mobility of the PEG chains is restricted by virtue of being covalently bonded to the inorganic LPSQ, this inhibits the crystallization of the PEG chains and thus reduces the crystallinity degree of the PEG. In consequence, the shorter the PEG chains, the higher the fraction of the restricted segments and the lower the crystallinity degree (Table 1).

Table 1. Thermal characteristics of PEG/PEO, PEG-grafted LPSQ, PEO-based and PMMA-based composites.

Entry	Sample	T_g (°C)	T_m (°C)	$\omega_{PEG\ (PEO)}$ [a]	ΔH_m ($J\ g^{-1}$)	X_c (%)
1	PEG1 [b]		38.7	1.00	166.8	85.5
2	PEG5 [b]		64.3	1.00	172.1	88.3
3	PEO [c]		65.6	1.00	146.7	75.2
4	PEG1-LPSQ (6)		36.5	0.80	66.3	42.5
5	PEG5-LPSQ (7)		57.3	0.61	83.2	69.9
6	PEO-Q1-2.5	−45.9	25.4 54.6	0.32 (0.60)	9.6 69.9	15.4 59.7
7	PEO-Q5-2.5	−51.2	57.9	0.10 (0.83)	142.1	78.4
8	PEO-Q5-1.0	−58.7	52.5	0.04 (0.93)	130.4	68.9
9	PMMA-Q1-2.5	−43.1 97.4		0.32	23.6	37.7
10	PMMA-Q5-2.5	−13.2		0.10	1.2	6.0

[a] weight fraction of PEG and PEO in the materials. [b] PEG1 and PEG5 are pristine PEG of molar masses of 1000 g mol^{-1} and 5000 g mol^{-1}. [c] PEO of molar mass of 100,000 g mol^{-1} used as matrix for the preparation of composites.

2.2. Elaboration and Characterization of Composite Materials Based on PEG-Grafted LPSQ

The PEG-grafted LPSQ hybrid materials were used as nanofiller and blended with poly(ethylene oxide) (PEO) or poly(methyl methacrylate) (PMMA) as matrices to elaborate different composites using the solvent casting process. Polymer nanocomposites can be defined as a class of materials where one or more phases with nanoscale dimensions are embedded in a polymer matrix [26]. Since the size of LPSQs are smaller than 4 nm (according to their molar mass), the resulting blends between PEG-grafted LPSQ and PEO or PMMA matrices can be qualified as polymer nanocomposites. Due to the small size of LPSQs, their low contrast with polymer matrix and their dilution (≤2.5 wt%) in the blends, it is difficult to analyze their dispersion and structure in the blends by WAXS as when carried out for LPSQs alone. Indeed, WAXS of blends containing PEG-grafted LPSQ and polyimide with a weight fraction of PEG-grafted LPSQ lower than 50 wt% did not show proof of the dispersion of functional LPSQ in the polyimide matrix [10].

2.2.1. DSC Analysis

The thermal behavior of the nanocomposite materials was investigated using DSC. The DSC traces were taken during the 2nd heating to remove any thermal history. Figures 2 and 3 show the heating thermograms for PEO matrix, PEG-grafted LPSQ and their composites. Table 1 summarizes glass transition temperature (T_g), T_m and X_c values obtained from the thermograms. As shown in Figures 2 and 3, a single T_g was observed for the resulting composites containing PEG1-LPSQ and PEG5-LPSQ, suggesting that PEO and PEG-grafted LPSQ are fully miscible in an amorphous phase. The high miscibility of the system is ascribed to the PEG-based LPSQ grafted chains having the same repeat unit structure as the PEO matrix. In the composites prepared with the same nanofiller e.g. PEG5-LPSQ, we observed a reduction in T_g with an increase in PEG5-LPSQ content (Table 1). This trend was also reported for nanocomposite electrolytes composing of PEO mixed with POSS functionalized with PEG of 637 g mol^{-1} [27]. This suggests that the grafted PEG chains in the LPSQ act as internal plasticizers to PEO. On the other hand, the composites containing the same weight fraction of silica (e.g., 2.5 wt%) but prepared with different PEG-grafted LPSQ didn't exhibit the same T_g. The PEO-Q1-2.5 composite shows a higher T_g (–45.9 °C) value than that of PEO-Q5-2.5 (–51.2 °C) composite (Table 1) suggesting that plasticizing character of short PEG chains (1000 g mol^{-1}) is less than that of long PEG chains (5000 g mol^{-1}).

Concerning the melting of PEO in the nanocomposites, as seen in Figures 2 and 3, both the T_m and melting enthalpy slightly decrease when PEG-LPSQ nanofillers are incorporated into PEO. Furthermore, we noted an additional small peak appears on the heating thermogram of the PEO-Q1-2.5 composite (Figure 2), indicating the formation of a new crystalline phase. This peak is very close to the peak of PEG5-LPSQ. This finding was already reported in the literature for the mixture of PEG-grafted POSS with a PEO matrix [27]. According to this paper, the concentration of LPSQ moieties in the new crystalline phase (LPSQ-rich phase) would be greater than that of the crystalline phase at a higher temperature (PEO-rich phase).

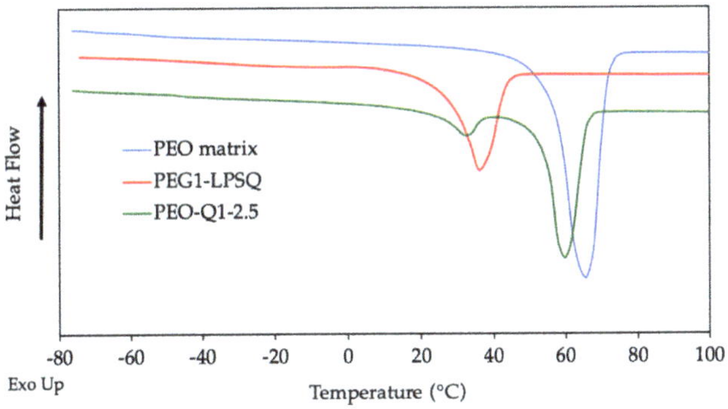

Figure 2. DSC thermograms during the second heating run for PEO matrix (blue solid line), PEG1-LPSQ (red solid line) and PEO-Q1-2.5 composite containing 2.5 wt% of silica (green solid line).

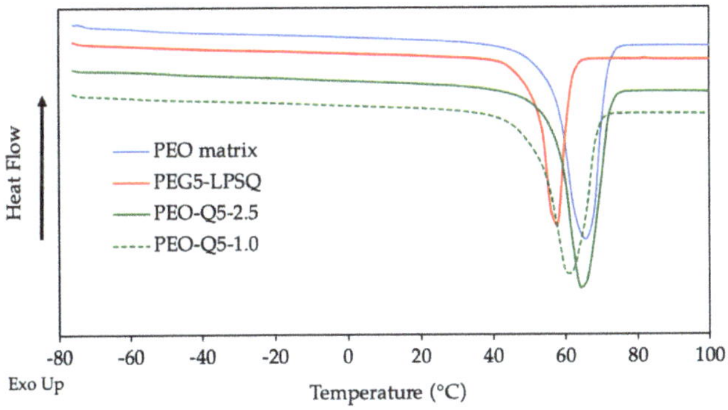

Figure 3. DSC thermograms during the second heating run for PEO matrix (blue solid line), PEG5-LPSQ (red solid line), PEO-Q5-2.5 composite containing 2.5 wt% of silica (green solid line) and PEO-Q5-1.0 composite containing 1 wt% of silica (green dash line).

The PEO/PMMA blends were largely studied in the literature because of their hydrophilic/hydrophobic character and their compatibility at certain compositions [16–19]. According to the work of Dionisio et al., PEO is miscible with PMMA up to a content of 25% of PEO in the blend [28]. The behavior of polymer blends, in particular the degree of miscibility, is strongly dependent on the molecular weight, the tacticity of PMMA and on the thermal history previous to the measurements. Decreases in both T_m of PEO and T_g of PMMA were generally observed for PEO/PMMA blends [29–31]. The DSC thermograms of PMMA-based composites and the PMMA matrix are shown in Figures S10–S12, and

DSC data are reported in Table 1. The PMMA-Q1-2.5 composite shows an endothermic melting peak at T_m = 28 °C corresponding to the crystalline phase of PEG (Figure S11) and two glass transition temperatures at −43 °C (PEG phase) and 97 °C (PMMA phase). Since the percentage of PEG in this composite is around 32 wt%, which is higher than that of the low limit of PEO content for a miscibility blend, the thermal behavior of the blend is mainly determined by the PEG component, and both T_m and X_c are less affected by the PMMA phase (Table 1, Entry 9). However, the T_g of the PMMA phase is lower than that of pristine PMMA (T_g = 117 °C) (Figure S10, Supplementary Materials), indicating the plasticization of PMMA by short PEG chains grafted on LPSQ. The PMMA-Q5-2.5 composite containing 10 wt% of grafted PEG shows a tiny melting peak at 47 °C and a unique T_g at −13 °C (Figure S12). According to the literature [15,16,18,28], at this PEO content, the blend is completely miscible showing a single T_g. The T_g observed for our sample is very far away from the expected T_g value (T_g = 89 °C) for a blend comprising 10 wt% of PEO and 90 wt% of PMMA [28]. Nevertheless, in our case, the PEO chains are grafted to the LPSQ structure, and this could greatly modify the behavior of polymer blends.

2.2.2. Rheology Analysis

Storage modulus (G′) and loss modulus (G″) are measured for all nanocomposites prepared with two different PEG-grafted LPSQs and their matrices (PEO and PMMA) by dynamic oscillatory measurements (Figures 4a,b and 5). The storage modulus of PEO-based nanocomposites presents a monotonic increase for all frequencies, including pure PEO. For pure PEO and its nanocomposites, the increment of frequency promotes the modulus because the polymer chains do not have enough time to relax at high frequencies. On the other hand, a low frequency or long time allows the relaxation of polymer chains to lower the modulus.

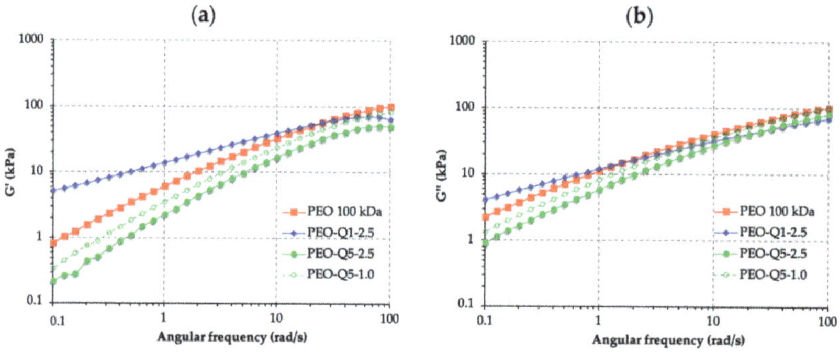

Figure 4. (**a**) Storage modulus (G′) and (**b**) loss modulus (G″) versus angular frequency of PEG-grafted LPSQ/PEO composites compared to neat PEO matrix of 100 kg mol^{-1}. Measurements were performed at 80 °C with a strain amplitude of 1%.

At 80 °C, pure PEO and PEO nanocomposites are under the melting state. As a consequence, the pure PEO melt exhibits a liquid-like behavior (G″ > G′) in all frequencies (Figure 4), while PEO composites present two different behaviors dependent on the type of functional LPSQ used. The PEO-Q1-2.5 nanocomposite prepared with the LPSQ functionalized with short PEG chains (1000 g mol^{-1}) shows solid-like behavior (G′ > G″) from 0.1 to 80 rad/s, while PEO-Q5-2.5 composite prepared with the LPSQ functionalized with longer PEG chains (5000 g mol^{-1}) shows a liquid-like behavior for all frequencies. In addition, we note that the values of storage modulus vary in the order G′ (PEO-Q1-2.5) > G′ (pure PEO) > G (PEO-Q5-1.0) > G′ (PEO-Q5-2.5) for almost all frequencies. These results suggest that PEG1-LPSQ is more efficient for the reinforcement of PEO matrix. Figure S13 (Supplementary Materials) shows the frequency dependence of tan δ = G″/G′ for nanoassociated (tan δ > 3), weakly associated (1 < tan δ < 3) and strongly associated (tan δ < 1) dispersed functional LPSQ molecules [32]. Here, we think that when the grafted PEG chains are short ($M_n < M_c$, where M_c of 4000 g mol^{-1} is the critical entanglement molecular weight of PEO) [33], the LPSQ molecules act as entangled attractors, resulting in higher entanglements near the surface of the LPSQ molecules, which reinforce the PEO matrix. On the contrary, when the grafted PEG chains are sufficiently long ($M_n > M_c$), grafted PEG chains act as entangled attractants, which lead to only PEG/PEO entanglements and PEG/PEO interactions, resulting in a plasticizing effect for the PEO matrix.

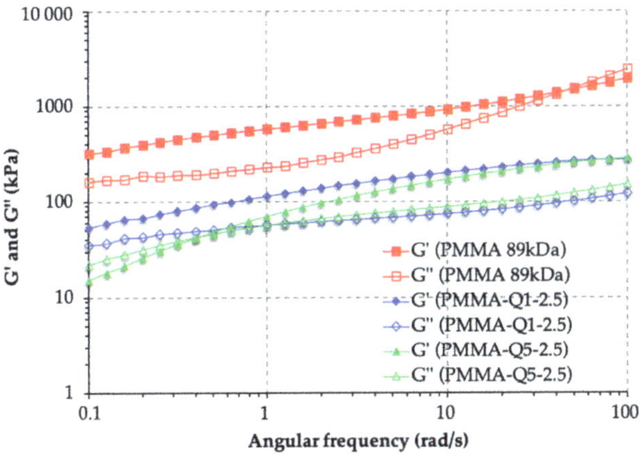

Figure 5. Storage (G′) and loss (G″) modulus versus angular frequency of PEG-grafted LPSQ/PMMA composites compared to neat PMMA matrix of 89 kg mol^{-1}. Measurements were performed at 150 °C with a strain amplitude of 1%.

It is obviously apparent from Figure 5 that the PMMA/PEG-LPSQ nanocomposites show a lower value of G′ and G″ compared to the pure PMMA. A decrease of one to one-and-a-haft decades in both the elastic (G′) and loss (G″) modulus lower than that of the pure PMMA is observed for all frequencies. The PMMA-Q1-2.5 composite exhibits a solid-like behavior for all frequencies, while the PMMA-Q5-2.5 composite is liquidlike at low frequencies. The transition from liquidlike to solid-like behavior occurs at the crossover frequency of G′ and G″ of 0.4 rad s^{-1}. This result illustrates that the dynamics of PMMA are significantly enhanced owing to the plasticizing nature of the PEG chains for the PMMA matrix. This effect increases with the grafted PEG chain length.

3. Materials and Methods

3.1. Materials

(3-Chloropropyl)trimethoxysilane (>97%), (3-Chloropropyl)triethoxysilane (95%), sodium azide (>99%), propargyl bromide (80% in toluene), sodium hydride (60% dispersion in oil), ethylenediaminetetraacetic acid (EDTA), poly(ethylene glycol) methyl ether of 1000 g mol^{-1} (PEG1) and 5000 g mol^{-1} (PEG5), poly(ethylene oxide) (PEO) of 100,000 g mol^{-1} and poly(methyl methacrylate) (PMMA) of 89,000 g mol^{-1} were purchased from Sigma-Aldrich (Saint Louis, MO, USA) and used as received. All solvents such as tetrahydrofuran (THF), dichloromethane (DCM), dimethyl sulfoxide (DMSO) and *N*,*N*-dimethyl formamide (DMF) and other reactants were purchased from commercial suppliers and used without further purification.

3.2. Characterization Techniques

Thermogravimetric measurements (TGA) were carried out on a 10 to 20 mg sample with a TGA Q500 apparatus (TA Instruments, New Castle, DE, USA) at a scan rate of 5 °C·min^{-1} from 40 to 800 °C under dynamic air atmosphere at a flow rate of 40 mL·min^{-1}.

Differential scanning calorimetry (DSC) experiments were carried out on a TA Instruments DSC Q20. The measurements were performed by heating around 10 mg of samples from −70 °C to 180 °C for hybrid material and commercial PEO. The heating rate was 10 °C min^{-1}. The melting temperature (T_m) and the enthalpies of melting ΔH_m were all extracted from the second heating scan. The degree of crystallinity (X_c) was calculated according to the following equation:

$$X_c = \frac{\Delta H_m}{w_{PEO} * \Delta H_m^o} * 100\% \quad (1)$$

where w_{PEO} is the weight fraction of PEO in the LPSQ hybrid materials, ΔH_m is the measured enthalpy of melting of pristine or grafted PEO and ΔH_m^o is the enthalpy of fusion of a 100% crystalline PEO taken as 195 J g^{-1} [34].

Wide-angle X-ray scattering (WAXS) experiments were performed on SAXSess-MC2 (Anton-Paar, GmbH, Vienna, Austria) with a sealed copper tube as the X-ray source (wavelength is 0.15417 nm (Cu K-α)) equipped with a Kratky block collimation system and using an image plate (IP 200 × 60 mm, pixel resolution 50 × 50 μm) as the detector. The system operates in line collimation, which allows achieving higher scattering intensities and images. Plates were read with the Cyclone (Perkin Elmer, Shelton, CT, USA) and Optiquant 3.0 software system. An exposure time of 30 min was long enough to provide a good signal-to-noise ratio (scattering angle range $0.1 < q < 28$ nm^{-1}). SAXSquant 3.5 was used as data analysis software.

Fourier-transform infrared (FTIR) analysis was performed on a Perkin Elmer Two FTIR Spectrometer with an ATR accessory.

All solid-state Cross Polarization Magic Angle Spinning (CPMAS) NMR spectra were obtained on a Bruker (Billerica, MA, USA) Avance-400 MHz NMR spectrometer operating at a ^{13}C and ^{29}Si resonance frequency of 101.6 MHz and 79.5 MHz, respectively. ^{13}C and ^{29}Si CPMAS experiments were performed with a commercial Bruker double-bearing probe. About 100 mg of samples was placed in zirconium dioxide rotors of 4 mm outer diameter and spun at a Magic Angle Spinning rate of 10 kHz. The CP technique [35] was applied with a ramped 1H pulse starting at 100% power and decreasing until 50% during the contact time in order to circumvent Hartmann–Hahn mismatches [36,37]. The contact times were 2 ms for ^{13}C CPMAS and 5 ms for ^{29}Si CPMAS. To improve the resolution, a dipolar decoupling GT8 pulse sequence was applied during the acquisition time [38]. To obtain a good signal-to-noise ratio, 6144 scans were accumulated using a delay of 2 s in the ^{13}C CPMAS experiment, and 4096 scans with a delay of 5 s in the ^{29}Si CPMAS experiment. The ^{13}C and ^{29}Si chemical shifts were referenced to tetramethylsilane.

Viscoelastic properties of the LPSQ hybrid or PEO and PMMA composite were studied using an Anton Paar Rheometer MCR 302 equipped with 8 mm diameter aluminum parallel disks. The polymer was pressed to a gap width of 0.9 mm. The complex shear modulus, $G^* = G' + iG''$, was then measured as a function of frequency by dynamically shearing the compound at a fixed strain of 1% over the frequency range 0.1–100 rad s^{-1} at 80 °C for the PEO composite and 150 °C for the PMMA composite. All measurements were performed in a dry air atmosphere.

Size exclusion chromatography (SEC) analysis was performed on an apparatus from Agilent Technologies (Santa Clara, CA, USA), which was equipped with a 1260 infinity pump, a 1260 infinity autosampler, a 1260 infinity UV photodiode array detector and a 1260 infinity RI detector. The stationary phase was composed of two PSS-SDV Linear M columns and a precolumn thermostated at 40 °C. Mobile phase was THF at 1 mL.min^{-1}. Samples were prepared at a concentration of 0.25 wt.% in THF containing 0.25 vol.% of toluene, as a flow marker. Injection volume was 20 μL. Polystyrene equivalent number-average and weight-average molar masses (M_n and M_w) and dispersities $Đ$ were calculated by means of PS calibration curve using PS-M Easivial (Agilent, USA).

3.3. *Synthesis of Chloropropyl Ladder PolySilsesQuioxane* (**2a**)

A solution of K_2CO_3 (20 mg, 0.144 mmol, 0.0036 eq) in water (2.6 mL) was added to a solution of (3-chloropropyl) triethoxysilane (9.6 g, 40 mmol, 1 eq) in THF (5.2 mL). Then, MeOH (800 μL) was added. The mixture was stirred at room temperature for 9 days. After removing THF under vacuum, the mixture was diluted with CH_2Cl_2 (40 mL) and water (10 mL). The organic phase was isolated and dried over $MgSO_4$; the solution was evaporated to give the final product **2a** (5 g, yield > 95%). ^{1}H NMR (300 MHz, $CDCl_3$) δ: 3.55 (broad peak, 2H, CH_2-Cl), 1.85 (broad peak, 2H, CH_2), 0.80 (broad peak, 2H, CH_2-Si).^{13}C NMR (75 MHz, $CDCl_3$) δ: 47.1 (CH_2-Cl), 26.5 (CH_2), 10.6 (broad, Si-CH_2). An amount of 4 mol% of ethoxy signal was still present, because of remaining unhydrolyzed functions. ^{29}Si CPMAS NMR: −58.3 (T^2), −68.5 (T^3). IR asymmetric stretching vibrations of the perpendicular bonds between the silicon and the oxygen of the LPSQ were at 1025.5 cm^{-1} and 1102.5 cm^{-1}. SEC analysis gave an M_n of 4 200 g mol^{-1} and a dispersity (Đ) of 2.48.

3.4. *Synthesis of Chloropropyl Ladder PolySilsesQuioxane* (**2b**)

A solution of K_2CO_3 (10 mg, 0.0725 mmol, 0.0036 eq) in water (1.3 mL) was added to a solution of (3-chloropropyl) trimethoxysilane (4 g, 20.129 mmol, 1eq) in THF (2.6 mL). The mixture was stirred at room temperature for 18 h. After removing THF under vacuum, the mixture was diluted with CH_2Cl_2 (30 mL) and water (10 mL). The aqueous phase was eliminated, and the organic phase was dried over $MgSO_4$ and filtered; the solution was evaporated to give the final product **2b** (668 mg, 26%). ^{1}H NMR (300 MHz, $CDCl_3$) δ: 3.54 (broad peak, 2H, CH_2-Cl), 1.85 (broad peak, 2H, CH_2), 0.80 (broad peak, 2H, CH_2-Si).^{13}C NMR (75 MHz, $CDCl_3$) δ: 47.0 (CH_2-Cl), 26.4 (CH_2), 9.9 (broad, Si-CH_2). Infrared asymmetric stretching vibrations of the perpendicular bonds between the silicon and the oxygen of the LPSQ were at 1028 cm^{-1} and 1099 cm^{-1}. SEC analysis gave a number-average molecular weight (M_n) of 3000 g mol^{-1} and a dispersity (Đ) of 2.7 based on polystyrene standards calibration. ^{29}Si CPMAS NMR: −58.3 (T^2), −68.5 (T^3).

3.5. *Synthesis of Azidopropyl Ladder PolySilsesQuioxane (N_3-LPSQ or* **3**)

NaN_3 (527 mg, 8.101 mmol, 1.5 eq) was added to a solution of **2a** (700 mg, 5.4 mmol, 1 eq) in DMF (7 mL). The reaction was stirred at 40 °C under an Argon atmosphere for 3 days. The mixture was diluted with CH_2Cl_2 (150 mL) and washed with water (400 mL). The aqueous phase was isolated and extracted with CH_2Cl_2 (20 mL × 5). The organic phases were combined and dried over $MgSO_4$; the solution was then evaporated to give the **N_3-LPSQ** (**3**) (631 mg, 4.64 mmol, 86%).

^{1}H NMR (300 MHz, $CDCl_3$) δ: 3.28 (broad peak, 2H, CH_2-N_3), 1.68 (broad peak, 2H, CH_2), 0.74–0.70 (broad peak, 2H, CH_2-Si). ^{13}C NMR (75 MHz, $CDCl_3$) δ: 53.5 (CH_2-N_3), 22.7 (CH_2), 9.6 (Si-CH_2). ^{29}Si CPMAS NMR: −57.5 (T^2), −67.7 (T^3). IR asymmetric stretching vibrations of the perpendicular bonds between the silicon and the oxygen of the LPSQ were the two bands at 1039 cm^{-1} and 1096 cm^{-1} and N_3 elongation band at 2090 cm^{-1}.

3.6. *Synthesis of Propargyl-Functionalized Poly (Ethylene Glycol) (PEG)*

- *Synthesis of propargyl-PEG of 1000 g mol^{-1}* (**4**)

In a 100 mL flask, polyethylene glycol monomethyl ether of 1000 g mol^{-1} (PEG1) (4.9 g, 4.9 mmol, 1 eq) was dissolved in anhydrous THF (4.9 mL) and cooled down to 0 °C. Under argon, the NaH 60% dispersed in oil (392 mg, 9.8 mmol, 2 eq) was introduced to the previous solution. The mixture was stirred at 0 °C for 1 h before adding in dropwise a solution of propargyl bromide 80% in toluene (2.73 mL, 24.5 mmol, 5 eq). After 24 h stirring at 30 °C, DCM was added, and the mixture was filtered. The filtrate was concentrated under vacuum; the resulting solid was crushed in cold ethyl ether, filtered and dried under vacuum. Propargyl-PEG1 (**4**) was obtained quantitatively as 5 g of a beige solid.

^{1}H NMR (300 MHz, $CDCl_3$) δ: 4.20 (m, 2H, propargyl CH_2), 3.67 (broad peak, 82H, OCH_2), 3.56 (m, 2H, OCH_2 next to OCH_3 end group), 3.39 (s, 3H, OCH_3), 2.45 (m, 1H, CH).

- *Synthesis of propargyl-PEG of 5000 g mol^{-1}* (**5**)

In a 100 mL flask, dried polyethylene glycol monomethyl ether of 5000 g mol^{-1} (PEG5) (10 g, 2 mmol, 1 eq) was dissolved in anhydrous THF (10 mL) at 60 °C and cooled down to 40 °C. Under argon, the NaH 60% dispersed in oil (120 mg, 3 mmol, 1.5 eq) was introduced slowly to the previous solution. The mixture was stirred at 40 °C for 1 h before adding in dropwise a solution of propargyl bromide 80% in toluene (664 µL, 6 mmol, 1.5 eq). After 48 h stirring at 40 °C, the warm mixture was filtered and rinsed with DCM. The filtrate was concentrated under vacuum and the concentrated solution was precipitated in cold diethyl ether. The solid was filtered and dried under vacuum. Propargyl-PEG5 (**5**) was obtained quantitatively as a beige solid.

^{1}H NMR (300 MHz, $CDCl_3$) δ: 4.19 (m, 2H, propargyl CH_2), 3.63 (m and broad peak, 452H, OCH_2), 3.36 (s, 3H, OCH_3), 2.43 (m, 1H, CH).

3.7. Grafting of PEG to the LPSQ by "Click Chemistry"

Two methods were used to graft PEG to the PLSQ. For the short PEG of 1000 g mol^{-1}, the click reaction was catalyzed by copper(I) while for the longer one (PEG of 5000 g mol^{-1}), the click reaction was carried out without catalyst at the melting state.

- *Click reaction catalyzed by copper (I)*—**PEG1-LPSQ** (**6**)

A solution of propargyl-PEG1 (**4**) (2.29 g, 2.205 mmol, 1 eq) in THF/H_2O (11.3 mL/5.6 mL) and LPSQ (300 mg, 2.205 mmol, 1 eq) was stirred at RT for 10 min. Then, $CuSO_4.5H_2O$ (275 mg, 1.102 mmol, 0.5 eq) and an aqueous solution of sodium ascorbate (218 mg, 1.102 mmol, 0.5 eq. in 3 mL H_2O) were added. The resulting mixture was stirred at RT. The reaction was monitored by ^{1}H NMR in $CDCl_3$ following the disappearance of the propargylic proton at 2.4 ppm.

After 3 days, because of the remaining residual propargyl signal, a second addition of a solution of sodium ascorbate (218 mg, 1.102 mmol, 0.5 eq in a minimal amount of water) and $CuSO_4.5H_2O$ (275 mg, 1.102 mmol, 0.5 eq) was performed. Two other portions of catalyst were required and added to the reactional mixture on the 6th and 9th day to reach the complete consummation of propargyl-PEG1.

After 11 days, THF was evaporated and the 2.5 g crude solid (green flakes) was dissolved in 15 mL of a mixture of acetone/water (2:1, *v/v*) under ultrasounds. This solution was then dialyzed against EDTA solution (400 mg L^{-1}) and deionized water to eliminate residual cupper ions and non-reacted LPSQ using tubing membrane with a cut off of 6000–8000 Da. After evaporation of water and drying in the vacuum, a beige solid was obtained called **PEG1-LPSQ** (**6**) (700 mg, 1.00 mmol, 45 wt%) as the single Z isomer. ^{13}C CPMAS NMR: (170 (spinning side bands)), 143.5 (Cq triazole), 123.1 (CH triazole), 70.9 (OCH_2, broad, PEG), 69.5 (CH_2), 57.5 (OCH_3), 50.9 (CH_2N), 22.9 (CH_2), 8.5 (CH_2-Si). ^{29}Si CPMAS NMR: −57.7 (T^2), −66.9 (T^3). FTIR: 2868 cm^{-1} (OCH_2 of PEG chain); N_3 function seems to be masked as only a very small peak was observed at 2089 cm^{-1} (Figure S6a, Supplementary Materials).

- *Click reaction in thermal condition without catalyst*—**PEG5-LPSQ** (**7**)

Propargyl-PEG5 (**5**) (741 mg, 0.147 mmol, 0.1 eq) was dried under vacuum for 3 h at 100 °C in order to eliminate water (PEG5 melts at 60 °C). Under argon atmosphere and under stirring, the **N_3-LPSQ** (**3**) (200 mg, 1.470 mmol, 1 eq) was added to the melt PEG. The reaction was monitored by ^{1}H NMR in $CDCl_3$ by following the disappearance of the propargylic proton signal of **5** located at 2.4 ppm. The reaction was carried out under stirring at 130 °C for 2 days. After cooling to RT, the solid crude product (940 mg) was dissolved in 15 mL of a mixture of acetone/water (2:1, *v/v*) under ultrasounds. This solution was then dialyzed against deionized water to eliminate non-reacted compounds using tubing membrane with a cut off of 25,000 Da. After evaporation of water and drying in the vacuum, a beige solid was obtained called **PEG5-LPSQ** (**7**) (800 mg, 85%wt). The NMR analysis indicates the formation of the 2 position isomers from cycloaddition 1–3 dipolar in 50/50 ratio. ^{13}C CPMAS NMR: (170 (spinning side bands) and 145.4 (Cq of the 2 triazole isomers), 124.7 and 134.5 (CH of the 2 triazole isomers), 71.0 (OCH_2, broad, PEG),

64.9 (CH_2), 61.5 (OCH_3), 54.2 (CH_2N), 23.5 (CH_2), 10.7 (CH_2-Si). ^{29}Si CPMAS NMR: −57.5 (T^2), −67.4 (T^3). FTIR: 2876.2 cm^{-1} (OCH_2 of PEG chain); N_3 elongation band at 2098 cm^{-1}, 1097 cm^{-1} and 1060 cm^{-1} for siloxane Si-O-Si.

*3.8. Preparation of Composite Materials from Hybrid Materials (PEG1-LPSQ (**6**) and PEG5-LPSQ (**7**)) Dispersed in 100K PEO and 89K PMMA Matrices*

Composites were prepared by the solvent casting process. They contained PEG-grafted LPSQ as the nanofiller and either PEO or PMMA as the matrices. The weight fraction of silica ($SiO_{1.5}$) in the composite was fixed at 2.5 wt% or 1 wt%. In a general protocol, the solutions of hybrid materials (**PEG1-LPSQ** (**6**) or **PEG5-LPSQ** (**7**) and polymer matrices (PEO or PMMA) were prepared separately in DCM or DMSO. The resulting solutions were then mixed together and stirred for 24 h. The solvent was evaporated at 60 °C under vacuum until a constant weight. Several samples were prepared, and the experimental data are summarized in Table 2. The resulting composites were labeled "*M-Qx-y*", where *M* corresponds to the polymer matrix (PEO or PMMA), *Qx* corresponds to the LPSQ functionalized with PEG, x = 1 or 5 corresponding to the grafted PEG of 1000 g mol^{-1} or 5000 g mol^{-1}, respectively, and y corresponds to the weight fraction of silica in the composite materials.

Table 2. Experimental data for the preparation of composite materials.

Composite	Nanofiller (mg)	$m_{nanofiller}$ (mg)	Polymer Matrix	m_{matrix} (mg)	wt% $SiO_{1.5}$ in Composite
PEO-Q1-2.5	PEG1-LPSQ (**6**)	51.6	PEO	78.4	2.5
PMMA-Q1-2.5	PEG1-LPSQ (**6**)	51.6	PMMA	78.4	2.5
PEO-Q5-1	PEG5-LPSQ (**7**)	10.1	PEO	139.1	1.0
PEO-Q5-2.5	PEG5-LPSQ (**7**)	25.3	PEO	124.7	2.5
PMMA-Q5-2.5	PEG5-LPSQ (**7**)	25.3	PMMA	124.7	2.5

4. Conclusions

In the present work, we have successfully synthesized two LPSQ molecules from either 3-chloropropyl triethoxysilane or 3-chloropropyl trimethoxysilane. According to the characterization results obtained by NMR, FTIR, WAXS and SEC, the LPSQ (**2a**) prepared from 3-chloropropyl triethoxysilane showed a better-defined structure than the other one because of the fast hydrolysis of the trimethoxysilane compound. The functionalization of LPSQ with PEG chains (M_n of 1000 g mol^{-1} and 5000 g mol^{-1}) was performed via click chemistry between azide groups of the functionalized LPSQ structure and propargyl ether on the PEG chains. Two click conditions were experimented, and both led to the expected PEG-grafted LPSQ hybrid materials. The latter were used as nanofiller to elaborate nanocomposite materials using either PEO or PMMA as matrix. In the case of PEO-based nanocomposites, for the same silica ($SiO_{1.5}$) content (e.g., 2.5 wt%), DSC analysis revealed that whatever the M_n of grafted PEG chains, T_m and crystallinity degree of the resulting nanocomposites shifted just a little, while rheology analysis showed a strong effect of M_n of grafted PEG chains on the viscoelastic properties of their nanocomposites. Indeed, PEG(1 kDa)-grafted LPSQ conferred a reinforcement character to the composite while PEG(5 kDa)-grafted LPSQ acted as a plasticizer, attested by the lower G′ and G″ modulus values of PEO-Q5-2.5 composite compared to those of PEO matrix. Consequently, if we are looking for the reinforcement effect of PEO matrix, it would be more relevant to use short-PEG-chain-grafted LPSQ than long-PEG-chain-grafted LPSQ. Concerning PMMA-based nanocomposites, two T_g values were observed for PMMA-Q1-2.5 composite with 32 wt% of PEG, indicating a partial miscible blend, while only one T_g value was obtained for PMMA-Q5-2.5 composite with 10 wt% of PEG, which is proof of a miscible blend. This observation is consistent with the literature results where the low limit of PEO content for a miscible

PEO/PMMA blend was around 25 wt%. The fact that the PEG chains are grafted does not change the miscibility behavior of PEG-grafted LPSQ/PMMA blends. However, the thermal properties such as T_g and T_m of PMMA-Q5-2.5 composite were strongly affected, as well as its viscoelastic properties. Indeed, the obtained T_g value (T_g = −13 °C) was far from the expected value (T_g around 90 °C) given in the literature [15,28], while G′ and G″ modulus values were one and a half decade lower than that of the pure PMMA matrix. The decrease in viscoelastic modulus of the PMMA-based composite is owing to the presence of PEG-grafted chains, which act as a plasticizer for PMMA. Albeit less pronounced, this effect was also observed for the partially miscible blend PMMA-Q1-2.5. The study of the thermodynamic and viscoelastic behavior of composites containing polymer-grafted LPSQ should constitute a fascinating field of investigation, while the PEO- or PMMA-based composites containing PEG-grafted LPSQ should lead to interesting solid electrolytes for lithium battery applications.

Supplementary Materials: The following supporting information can be downloaded at: https://www.mdpi.com/article/10.3390/molecules29245832/s1, Figure S1: ^{29}Si spectrum of LPSQ **2a**; Figure S2: ^{29}Si spectrum of LPSQ **2b**; Figure S3: infrared spectrum of LPSQ **2a**; Figure S4: TGA thermogram of PEG1-LPSQ (**6**) and associated calculation of the click reaction yield; Figure S5: TGA thermogram of PEG5-LPSQ (**7**) and associated calculation of the click reaction yield; Figure S6: FTIR spectra of (a) PEG1-LPSQ (**6**) and (b) PEG5-LPSQ (**7**); Figure S7: ^{13}C NMR spectra of (a) PEG1-LPSQ (**6**) and (b) PEG5-LPSQ (**7**); Figure S8: DSC thermograms during the second heating run of pristine PEG1 and PEG1-grafted LPSQ (**6**); Figure S9: DSC thermograms during the second heating run of pristine PEG5 and PEG5-grafted LPSQ (**7**); Figure S10: DSC thermogram during the second heating run of PMMA matrix of 89,000 g mol^{-1}; Figure S11: DSC thermogram during the second heating run of PMMA-Q1-2.5 composite; Figure S12: DSC thermogram during the second heating run of PMMA-Q5-2.5 composite; Figure S13: Tan δ versus frequency of pure PEO 100 kDa and its nanocomposites.

Author Contributions: Conceptualization, E.B. (Eric Besson), S.G. and T.N.T.P.; methodology, D.M., E.B. (Eric Besson) and T.N.T.P.; formal analysis, D.M., M.R., E.B. (Emily Bloch) and T.N.T.P.; investigation, D.M., M.R., E.B. (Eric Besson) and T.N.T.P.; data curation, D.M., M.R., E.B. (Eric Besson), S.G. and T.N.T.P.; writing—original draft preparation: D.M., E.B. (Eric Besson), S.G. and T.N.T.P.; writing—review and editing, D.M., E.B. (Eric Besson), S.G. and T.N.T.P.; supervision, E.B. (Eric Besson), S.G. and T.N.T.P. All authors have read and agreed to the published version of the manuscript.

Funding: This research was funded by Aix-Marseille university and CNRS.

Institutional Review Board Statement: Not applicable.

Informed Consent Statement: Not applicable.

Data Availability Statement: The data that support the findings of this study are available from the corresponding author upon reasonable request.

Conflicts of Interest: The authors declare no conflicts of interest.

References

1. Usuki, A.; Kawasumi, M.; Kojima, Y.; Okada, A.; Kurauchi, T.; Kamigaito, O. Swelling Behavior of Montmorillonite Cation Exchanged for ω-Amino Acids by ∈-Caprolactam. *J. Mater. Res.* **1993**, *8*, 1174–1178. [CrossRef]
2. Urtekin, G.; Ullah, M.S.; Yildirim, R.; Ozkoc, G.; Kodal, M. A Comprehensive Review of the Recent Developments in Thermoplastics and Rubber Blends-based Composites and Nanocomposites. *Polym. Compos.* **2023**, *44*, 8303–8329. [CrossRef]
3. Idumah, C.I.; Obele, C.M. Understanding Interfacial Influence on Properties of Polymer Nanocomposites. *Surf. Interfaces* **2021**, *22*, 100879. [CrossRef]
4. Zhang, C.; Ma, C.; Yao, X.; Yang, X.; Zhang, W.; Zhang, G. Transparent and Flexible Vesicles/Lamellae Mixed Matrix Organic-Bridged Polysilsesquioxane Coatings as Water Vapor Barrier. *Prog. Org. Coat.* **2023**, *174*, 107236. [CrossRef]
5. Mohamed, M.G.; Kuo, S.-W. Progress in the Self-Assembly of Organic/Inorganic Polyhedral Oligomeric Silsesquioxane (POSS) Hybrids. *Soft Matter* **2022**, *18*, 5535–5561. [CrossRef]
6. Kajihara, K.; Kanamori, K.; Shimojima, A. Current Status of Sol–Gel Processing of Glasses, Ceramics, and Organic–Inorganic Hybrids: A Brief Review. *J. Ceram. Soc. Jpn.* **2022**, *130*, 575–583. [CrossRef]

7. Chen, F.; Lin, F.; Zhang, Q.; Cai, R.; Wu, Y.; Ma, X. Polyhedral Oligomeric Silsesquioxane Hybrid Polymers: Well-Defined Architectural Design and Potential Functional Applications. *Macromol. Rapid Commun.* **2019**, *40*, 1900101. [CrossRef]
8. Sunchu, P.; Besson, E.; Phan, T.N.T.; Gastaldi, S. Dual Role of Alkoxyamine-Functionalized Lamellar Materials in the Synthesis of Hybrid Polymers. *Eur. Polym. J.* **2024**, *211*, 112956. [CrossRef]
9. Kim, J.; Park, Y.; Kwon, M.S. Recent Progress in Ladder-like Polysilsesquioxane: Synthesis and Applications. *Mater. Chem. Front.* **2024**, *8*, 2689–2726. [CrossRef]
10. Hwang, S.O.; Lee, J.Y.; Lee, J.-H. Effect of the Silsesquioxane Structure on the Mechanical Properties of the Silsesquioxane-Reinforced Polymer Composite Films. *Prog. Org. Coat.* **2019**, *137*, 105316. [CrossRef]
11. Kim, R.I.; Shin, J.H.; Lee, J.S.; Lee, J.-H.; Lee, A.S.; Hwang, S.S. Tunable Crystalline Phases in UV-Curable PEG-Grafted Ladder-Structured Silsesquioxane/Polyimide Composites. *Materials* **2020**, *13*, 2295. [CrossRef] [PubMed]
12. Kumar, S.K.; Benicewicz, B.C.; Vaia, R.A.; Winey, K.I. 50th Anniversary Perspective: Are Polymer Nanocomposites Practical for Applications? *Macromolecules* **2017**, *50*, 714–731. [CrossRef]
13. Mauger, A.; Julien, C.M.; Paolella, A.; Armand, M.; Zaghib, K. Building Better Batteries in the Solid State: A Review. *Materials* **2019**, *12*, 3892. [CrossRef]
14. Brodeck, M.; Alvarez, F.; Colmenero, J.; Richter, D. Single Chain Dynamic Structure Factor of Poly(Ethylene Oxide) in Dynamically Asymmetric Blends with Poly(Methyl Methacrylate). Neutron Scattering and Molecular Dynamics Simulations. *Macromolecules* **2012**, *45*, 536–542. [CrossRef]
15. Liau, W.B.; Chang, C.F. Casting Solvent Effect on Crystallization Behavior and Morphology of Poly(Ethylene Oxide)/Poly(Methyl Methacrylate). *J. Appl. Polym. Sci.* **2000**, *76*, 1627–1636. [CrossRef]
16. Liu, J.; Sakai, V.G.; Maranas, J.K. Composition Dependence of Segmental Dynamics of Poly(Methyl Methacrylate) in Miscible Blends with Poly(Ethylene Oxide). *Macromolecules* **2006**, *39*, 2866–2874. [CrossRef]
17. Ghelichi, M.; Qazvini, N.T.; Jafari, S.H.; Khonakdar, H.A.; Farajollahi, Y.; Scheffler, C. Conformational, Thermal, and Ionic Conductivity Behavior of PEO in PEO/PMMA Miscible Blend: Investigating the Effect of Lithium Salt. *J. Appl. Polym. Sci.* **2013**, *129*, 1868–1874. [CrossRef]
18. García Sakai, V.; Maranas, J.K.; Peral, I.; Copley, J.R.D. Dynamics of PEO in Blends with PMMA: Study of the Effects of Blend Composition via Quasi-Elastic Neutron Scattering. *Macromolecules* **2008**, *41*, 3701–3710. [CrossRef]
19. Luo, C.; Chen, W.; Gao, Y. Feather-like Morphology of Poly(Methyl Methacrylate)/Poly(Ethylene Oxide) Blends: The Effect of Cooling Rate and Poly(Methyl Methacrylate) Content. *J. Appl. Polym. Sci.* **2015**, *132*, app.41705. [CrossRef]
20. Kaneko, Y.; Toyodome, H.; Shoiriki, M.; Iyi, N. Preparation of Ionic Silsesquioxanes with Regular Structures and Their Hybridization. *Int. J. Polym. Sci.* **2012**, *2012*, 684278. [CrossRef]
21. Lee, A.S.S.; Choi, S.-S.; Jang, S.-H.; Hwang, S.S.; Baek, K.-Y. Preparation of High Modulus Thin Films Based on Photocurable Azido-Functionalized Ladder-like Structured Polysilsesquioxanes. *Macromol. Res.* **2014**, *22*, 1109–1114. [CrossRef]
22. Park, E.S.; Ro, H.W.; Nguyen, C.V.; Jaffe, R.L.; Yoon, D.Y. Infrared Spectroscopy Study of Microstructures of Poly(Silsesquioxane)s. *Chem. Mater.* **2008**, *20*, 1548–1554. [CrossRef]
23. Liu, F.; Zeng, X.; Lai, X.; Li, H. Synthesis and Characterization of Polyphenylsilsesquioxane Terminated with Methyl and Vinyl Groups Low-Melting Glass. *J. Adhes. Sci. Technol.* **2017**, *31*, 2399–2409. [CrossRef]
24. Malvi, B.; Sarkar, B.R.; Pati, D.; Mathew, R.; Ajithkumar, T.G.; Sen Gupta, S. "Clickable" SBA-15 Mesoporous Materials: Synthesis, Characterization and Their Reaction with Alkynes. *J. Mater. Chem.* **2009**, *19*, 1409–1416. [CrossRef]
25. Rambarran, T.; Gonzaga, F.; Brook, M.A. Generic, Metal-Free Cross-Linking and Modification of Silicone Elastomers Using Click Ligation. *Macromolecules* **2012**, *45*, 2276–2285. [CrossRef]
26. Reduwan Billah, S.M. Composites and Nanocomposites. In *Cellulose-Based Superabsorbent Hydrogels*; Mondal, M.I.H., Ed.; Polymers and Polymeric Composites: A Reference Series; Springer International Publishing: Cham, Switzerland, 2019; pp. 1–67, ISBN 978-3-319-76573-0.
27. Jung, G.; Choi, J.H.; Lee, J.K. Thermal Behavior and Ion Conductivity of Polyethylene Oxide/Polyhedral Oligomeric Silsesquioxane Nanocomposite Electrolytes. *Adv. Polym. Technol.* **2015**, *34*, adv.21499. [CrossRef]
28. Dionísio, M.; Fernandes, A.C.; Mano, J.F.; Correia, N.T.; Sousa, R.C. Relaxation Studies in PEO/PMMA Blends. *Macromolecules* **2000**, *33*, 1002–1011. [CrossRef]
29. Cortazar, M.M.; Calahorra, M.E.; Guzmán, G.M. Melting Point Depression in Poly (Ethylene Oxide)-Poly (Methyl Methacrylate) Blends. *Eur. Polym. J.* **1982**, *18*, 165–166. [CrossRef]
30. Li, X.; Hsu, S.L. An Analysis of the Crystallization Behavior of Poly(Ethylene Oxide)/Poly(Methyl Methacrylate) Blends by Spectroscopic and Calorimetric Techniques. *J. Polym. Sci. Polym. Phys. Ed.* **1984**, *22*, 1331–1342. [CrossRef]
31. Martuscelli, E.; Canetti, M.; Vicini, L.; Seves, A. First Results of Small and Wide Angle X-Ray Scattering of Poly(Ethylene Oxide)/Poly(Methyl Methacrylate) Binary Blends. *Polymer* **1982**, *23*, 331–334. [CrossRef]
32. Hyun, Y.H.; Lim, S.T.; Choi, H.J.; Jhon, M.S. Rheology of Poly(Ethylene Oxide)/Organoclay Nanocomposites. *Macromolecules* **2001**, *34*, 8084–8093. [CrossRef]
33. Anderson, B.J.; Zukoski, C.F. Rheology and Microstructure of Polymer Nanocomposite Melts: Variation of Polymer Segment–Surface Interaction. *Langmuir* **2010**, *26*, 8709–8720. [CrossRef] [PubMed]
34. Griffin, L.O. *Physical Constants of Linerar Homopolymers*; Springer: Berlin/Heidelberg, Germany, 1986.

35. Schaefer, J.; Stejskal, E.O. Carbon-13 Nuclear Magnetic Resonance of Polymers Spinning at the Magic Angle. *J. Am. Chem. Soc.* **1976**, *98*, 1031–1032. [CrossRef]
36. Cook, R.L.; Langford, C.H.; Yamdagni, R.; Preston, C.M. A Modified Cross-Polarization Magic Angle Spinning 13C NMR Procedure for the Study of Humic Materials. *Anal. Chem.* **1996**, *68*, 3979–3986. [CrossRef]
37. Peersen, O.B.; Wu, X.L.; Kustanovich, I.; Smith, S.O. Variable-Amplitude Cross-Polarization MAS NMR. *J. Magn. Reson. A* **1993**, *104*, 334–339. [CrossRef]
38. Gerbaud, G.; Ziarelli, F.; Caldarelli, S. Increasing the Robustness of Heteronuclear Decoupling in Magic-Angle Sample Spinning Solid-State NMR. *Chem. Phys. Lett.* **2003**, *377*, 1–5. [CrossRef]

Disclaimer/Publisher's Note: The statements, opinions and data contained in all publications are solely those of the individual author(s) and contributor(s) and not of MDPI and/or the editor(s). MDPI and/or the editor(s) disclaim responsibility for any injury to people or property resulting from any ideas, methods, instructions or products referred to in the content.

molecules

MDPI

Article

Optical Properties and Antimicrobial Activity of Si/PVP Hybrid Material Combined with Antibiotics

Lilia Yordanova [1], Yoanna Kostova [2], Elitsa Pavlova [3], Albena Bachvarova-Nedelcheva [4,*], Iliana Ivanova [1] and Elena Nenova [1]

1 Faculty of Biology, Sofia University "St. Kliment Ohridski", 8 Dragan Tsankov Blvd., 1164 Sofia, Bulgaria; lilia_petrova3@abv.bg (L.Y.); iaivanova@biofac.uni-sofia.bg (I.I.); nenova@uni-sofia.bg (E.N.)

2 Institute of Metal Science, Equipment and Technologies with Hydro- and Aerodynamics Centre "Acad. A. Balevski", Bulgarian Academy of Sciences, Shipchenski Prohod Str., 67, 1574 Sofia, Bulgaria; y_kostova@ims.bas.bg

3 Faculty of Physics, Sofia University "St. Kliment Ohridski", 5 James Boucher Blvd., 1164 Sofia, Bulgaria; elli_pavlova@abv.bg

4 Institute of General and Inorganic Chemistry, Bulgarian Academy of Sciences, Acad. G. Bonchev Str., Bl. 11, 1113 Sofia, Bulgaria

* Correspondence: albenadb@svr.igic.bas.bg

Abstract: Silica–poly (vinylpyrrolidone) hybrid material was prepared using the sol–gel method. Tetramethyl ortosilane (TMOS) was used as a silica precursor. XRD analysis established that the as-prepared material is amorphous. The morphological structure of the final product was determined by the incorporated PVP. The UV–Vis analysis showed that the obtained hybrid exhibited absorption in the ultraviolet range. The antimicrobial activity of the SiO_2/15PVP hybrid material was tested on *Staphylococcus epidermidis* ATCC 14990, *Salmonella typhimurium* ATCC BAA-2162, *Candida albicans*, and *Saccharomyces cerevisiae* in combination with the following antibiotics: Vancomycin for Gram-positive bacteria, Ciprofloxacin for Gram-negative bacteria, and Nystatin for yeast. The results confirmed a concentration-dependent synergistic effect of the antibiotic in combination with the TM15/PVP hybrid particles, especially at their highest concentration of 100 mg/mL on Gram-positive bacteria and for the Gram-negative *Salmonella*. On *Candida albicans* ATCC 18804 and *Saccharomyces cerevisiae* CCY 21-6-3, the effect was synergistic again, and a fungicidal effect was observed at 6.25 and 1.50 mg/mL for the antibiotic concentration and concentrations of hybrid material at 100 mg/mL. The toxicity on *Daphnia magna* was also tested. The registered prooxidant activity of SiO_2/15PVP shows possible applications at very low concentrations. The obtained results demonstrate the possibility of clinical implementations of the newly synthesized hybrid material.

Keywords: hybrid; optical properties; antibacterial properties; antibiotics; prooxidant properties; environmental toxicity control

1. Introduction

One of the important extensions of the sol–gel process is the preparation of a group of transparent amorphous materials consisting of an organic polymer and a metal oxide such as silica gel [1]. These hybrid materials are constituted by an amorphous network of inorganic and organic phases with strong interface interactions, typically through coupling agent molecules. Therefore, by choosing the appropriate phases, hybrid materials with desirable properties can be designed for multiple purposes [2,3].

It is recognized that nanoparticles, with their unique properties, such as their high surface area to volume ratio, quantum effects, enhanced reactivity, antibacterial activity, oxidation resistance, and high thermal conductivity, are leading the way in the world of modern science and technology toward a true revolution in various fields, including medicine, electronics, and environmental protection [3]. Among the wide range of nanomaterials, mesoporous nanoparticles stand out with exceptional properties such as biocompatibility, easy modification, and the ability to accommodate drug molecules [4]. Therefore, these kinds of materials are excellent candidates for the growing problem of bacterial antibiotic resistance worldwide [5].

Generally, silica nanomaterials are still of great interest to modern science and biomedicine due to their excellent properties that enable numerous biological applications [6]. Inorganic silica nanoparticles have a high surface-to-volume ratio and exhibit tissue compatibility and stability in biological macro-organisms. Due to its pronounced biocompatibility, inertness, low toxicity, resistance to acidic conditions, and thermal resistance, silica has the ability to enter into a variety of biochemical interactions due to its ability to form covalent bonds with macromolecules, such as nucleic acids, antibodies, fluorescent molecules, thanks to the silanol groups in its structure [7]. Silica nanoparticles are highly prevalent in nanotechnology because their production is an easy and cheap process. They are also used as biocompatible pharmaceutical additives. Numerous studies have been developed that prove the low toxicity and high bactericidal efficiency of mesoporous silica nanoparticles whose surface is modified with organic polymers, for example, PVP, and whose interior is loaded with antibiotics [8].

It is well known that the generation of reactive oxygen species (ROS) is one of the main toxicity mechanisms by which nanoparticles inhibit bacterial growth. The generation of ROS is an important biomarker regarding the redox properties of substances. By following the kinetics of the generation of free radicals and ROS, the properties of newly synthesized materials can be evaluated as well as their effects on the cascade of reactions causing the formation and accumulation of antibacterial free radicals and ROS [9,10].

Chemiluminescence analysis is a rapid and sensitive method for such studies. It is applied to follow the dynamics of free radical and ROS generation and to determine the prooxidant/antioxidant activity of various materials. The responses are recorded in the range of 480–580 nm and can be used to estimate the quantum yield [10–12]. The redox activity of the newly synthesized material was tested in Fenton's model chemical reaction system, generating free radicals and ROS, at 7.4/37 °C (physiological), using the activated chemiluminescence method [11]. In general, the higher the signal, the more ROS are formed and the pro-oxidant effect is demonstrated. On the contrary, if the signal is below the control level, antioxidant activity can be found.

The combination of nanoparticles (NPs) with antibiotics may improve the antibacterial effect through additive or synergistic antibacterial effects by chemically or physically weakening the bacteria [13,14]. It has been shown that such mechanisms weaken bacteria by the creation of reactive oxygen species, antibacterial ions, or elevated temperatures [15]. The synergistic effects between metal and metal oxide NPs (MNPs) and commercial antimicrobial drugs have been studied for several years [14]. Most of the synergistic studies focus on silver nanoparticles (AgNPs), gold (Au), copper (Cu), copper oxide (CuO), copper sulfide (CuS), iron (Fe), iron oxide (Fe_3O_4/Fe_2O_3), zinc, zinc oxide (ZnO), and platinum (Pt). MNPs have been combined with several antibiotic, antifungal, and antiviral agents, but a high number of compounds still remain unexplored. However, their mechanism of action is still not completely understood [16]. Recently, it has been reported that the conjugation of MNPs with other antimicrobial compounds may enhance their effectiveness [14]. Some authors have found new approaches in the fight against pathogens including the revival of old antibiotics, to overcome the current drug resistance emergency [13,17]. Drug combination is a common strategy in clinical practice, and its therapeutic success has been found in the treatment of acquired immunodeficiency syndrome (AIDS), cancer, cardiovascular disease, and microbial infections [14].

The present work is a continuation of our previous investigations on the synthesis and antibacterial properties of sol–gel-derived hybrid silica–poly (vinylpyrrolidone) materials [8]. This study particularly demonstrates that it is possible to find effective combinations of nanostructures and conventional antibiotics to prevent the formation of bacterial biofilms and bacterial resistance. It was discovered that in combination, nanoparticles and antibiotics exert significant bactericidal activity against both susceptible and already resistant strains. The synergistic antimicrobial effect that allows for the inhibition of bacterial growth, using low concentrations of both the antibiotic and the nanoparticles, emphasizes the novelty of the paper.

2. Results and Discussion

2.1. XRD and SEM Observations

Figure 1 displays the X-ray diffraction patterns of samples with 15% PVP content as well as pure TMOS and PVP samples. The figure illustrates that all gels are amorphous. This result correlates well with those obtained in our previous investigations [8].

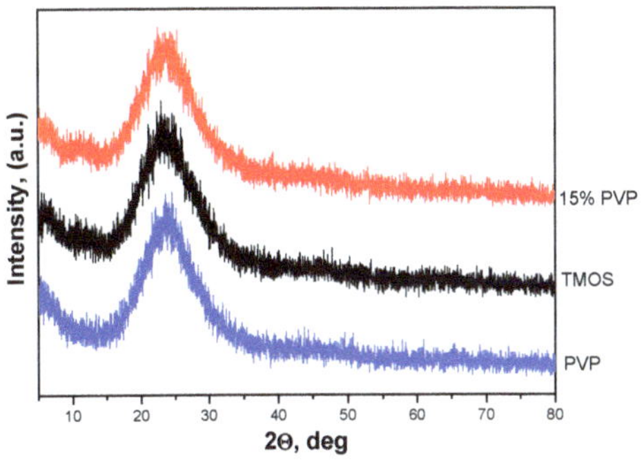

Figure 1. XRD patterns of SiO_2/15PVP, TMOS, and PVP.

SEM images of the investigated gel are shown in Figure 2a,b. The analysis of the morphology of the SiO_2/15PVP sol–gel derived powder showed clusters of agglomerated particles. The same images also exhibited the presence of particles with irregular shapes. However, other samples with larger quantities of PVP are characterized by such particle agglomerates [18]. This is due to the high linking ability of PVP. Some authors [18] have stated that during the formation of the silica core after the hydrolysis and polycondensation of TEOS and PVP, molecules interact with the silica particles via electrostatic and intermolecular hydrogen bonds between the OH groups of silica and C=O groups of PVP. Consequently, at higher PVP (>10 mol%) concentrations, large particle clusters could be formed during the sol–gel process. This influence of PVP on morphology is in accordance with our previous study [8].

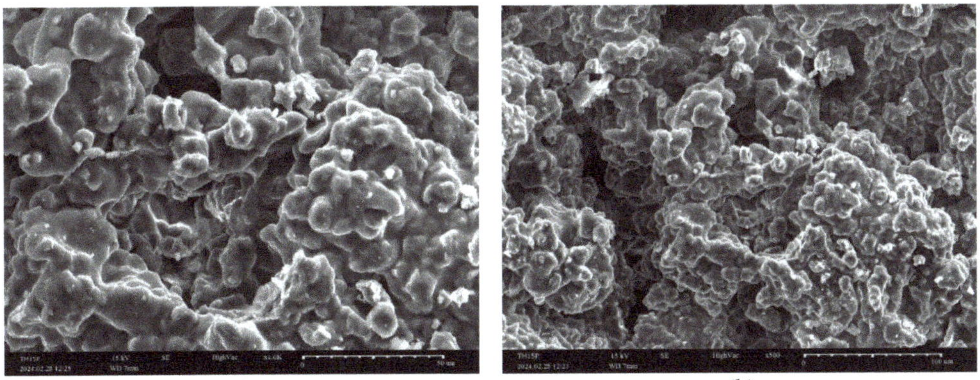

(a) (b)

Figure 2. SEM micrographs of a sample with 15% PVP at different magnifications (**a**,**b**).

2.2. UV–Vis Characterization

Aiming to analyze the optical properties of the produced gel, UV–Vis spectra were applied in the range from 190 to 500 nm for pure TMOS, PVP, and the obtained hybrid material (Figure 3). As can be seen from the figure, all spectra exhibited bands in the UV region only. Several absorption peaks (215, 225, and 235 nm) are characteristic of pure PVP, with that at 225 nm being the most intensive one. The increased absorption of PVP in the UV areas is probably due to the type of transitions $n \rightarrow \pi^*$ occurring between pairs of free charge carriers derived from oxygen atoms and vacant states [19]. TMOS showed strong absorption for ultraviolet wavelengths below 230 nm, which is consistent with the results presented by Xu et al. [20]. This phenomenon can be caused by defects in the surface structure of the produced nanoparticles, resulting in the formation of an incomplete four-wall network of Si–O–Si, which is characterized by local optical activity only [20]. The spectra of the obtained hybrid SiO_2/15PVP are characterized by a sharp absorption edge at 220 nm. In addition, an increase in the absorption of the composite in comparison to the pure TMOS was recorded. Using the registered UV–Vis spectra, the cut-off and energy gaps (Eg) values are shown in Table 1. The calculated Eg values for pure amorphous precursors TMOS and PVP are 5.43 and 4.63 eV, respectively, which corresponds well with the data obtained by other authors [19,21]. It has to be noted that the Eg value of pure TMOS is usually found to be above 9 eV, as reported by Weinberg et al. [21]. Nowadays, the production of SiO_2 nanostructures with lower Eg values can significantly affect the performance improvement of electronic components used in production processes that employ silica [19]. Looking back to Table 1, it can be seen that the band gap of the hybrid is 5.22 eV, with it shifting to lower energy waves, probably due to the nanometric size of the obtained gel. Bearing in mind that the absorption of the hybrid increases in comparison to the pure TMOS, it might be suggested that the obtained material could be used as an efficient absorber of ultraviolet radiation.

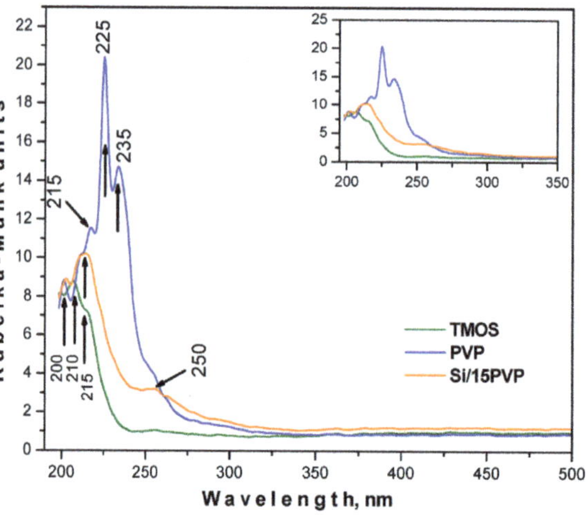

Figure 3. UV–Vis spectra for the pure TMOS, PVP, and SiO_2/15PVP hybrid. The inset shows the spectra in the range of 200–350 nm.

Table 1. Cut-off and optical band gap values for the investigated materials.

Sample	Cut-Off, nm	Eg, eV
PVP	267.77	4.63
TMOS	228.16	5.43
SiO_2/15PVP	237.61	5.22

2.3. AFM Investigations

The AFM images in Figure 4 show the morphology of the micro- and nanostructured surface of the hybrid SiO_2/15PVP material. The scanned area has dimensions of $5 \times 5\ \mu m^2$ and a scanning speed of 10 kH. From Figure 4a, it can be seen that the sample has a relatively smooth surface with negligible irregularities but significant homogeneity. Figure 4b is a three-dimensional topographic map from which the main parameters of the AFM analysis are determined. The inset in Figure 4b exhibits the hysteresis of the height distribution profiles of surfaces from which the mean square roughness (RMS) is determined. It represents the standard deviation of the height value in the selected region, which is 0.089 nm, and the calculated Ra (roughness average) for the scanned area is 0.055 nm. The surface height distribution profile (Figure 4c) obtained from the 3D image shows that the surface morphology of the inorganic–organic hybrid SiO_2/15PVP material is homogeneous and structured by minor irregularities with relatively low height. Other AFM studies of sol–gel hybrid materials show a good homogeneous continuous structure, as well as certain aspects of the morphology of the surface depending on the chemical composition and percentage of the organic additive [22,23].

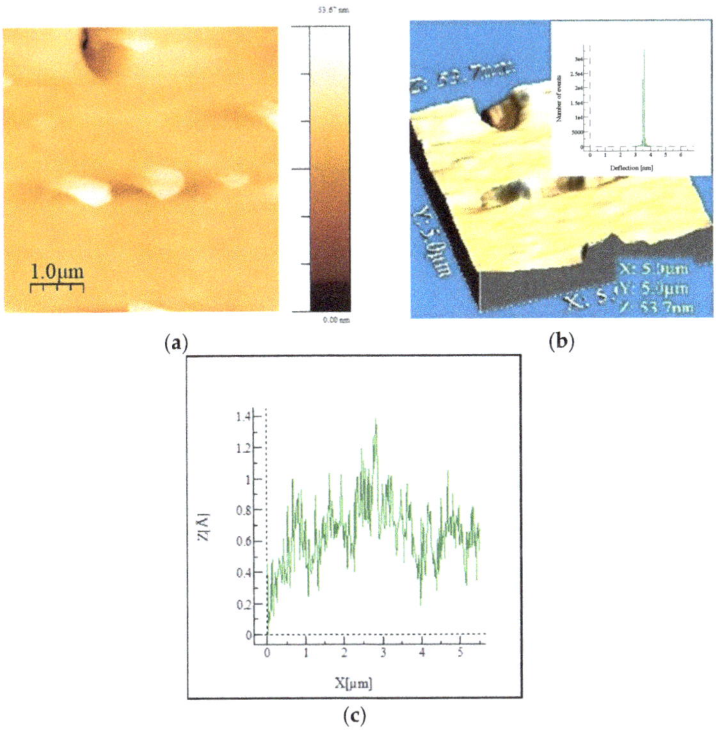

Figure 4. AFM images of the investigated SiO_2/15PVP hybrid: 2D (**a**) and 3D (**b**) surface topography and roughness profile (**c**) of the material.

2.4. Antibacterial Properties

In our research, we tested the effect of SiO_2/15PVP hybrid nanomaterials alone and in combination with popular antibiotics: Vancomycin against the Gram-positive bacteria *Staphylococcus epidermidis* ATCC 14990, Ciprofloxacin against the Gram-negative bacteria *Salmonella typhimurium* ATCC BAA-2162, and Nystatin against two eukaryotic cells—the yeasts *Candida albicans* ATCC 18804 *and Saccharomyces cerevisiae* CCY 21-6-3.

Figure 5 shows the amount (in CFU/mL) of *Staphylococcus epidermidis* ATCC 14990 at different concentrations of SiO_2/15PVP (100 mg/mL, 50 mg/mL, and 25 mg/mL) in combination with the antibiotic Vancomycin and compared to the effect of the antibiotic alone. The MBC of the single action of the antibiotic Vancomycin against *Staphylococcus epidermidis* was determined—125 µg/mL, while in the combination of Vancomycin with the highest concentration of SiO_2/15PVP (100 mg/mL), the MBC already decreased to only 62.5 µg/mL for the antibiotic concentration. The results in Figure 5 show that the minimum inhibition concentration (MIC) of single Vancomycin is 62.5 µg/mL, and the minimum inhibition concentration of 100 mg/mL SiO_2/15PVP in combination with Vancomycin is 31.5 µg/mL.

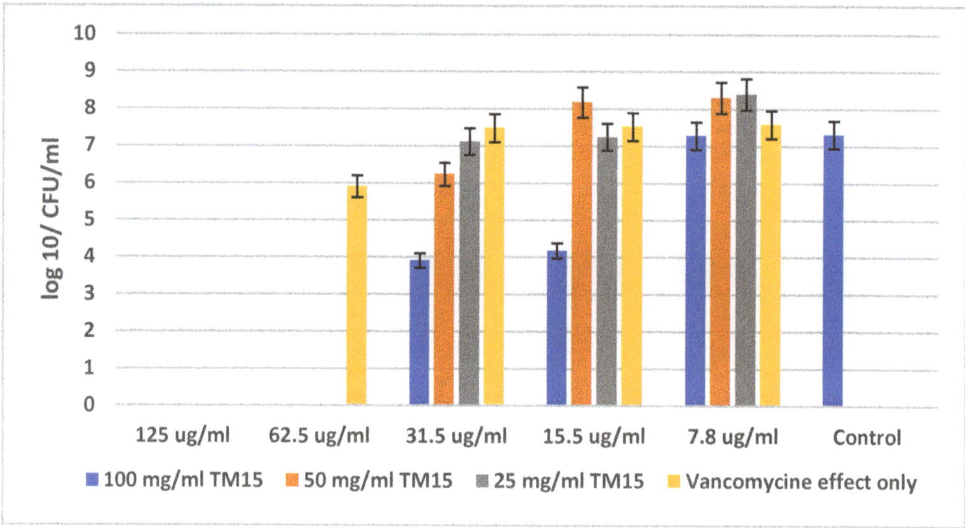

Figure 5. Effect of SiO_2/15PVP in combination with Vancomycin against *Staphylococcus epidermidis* ATCC 14990 at 24 h of treatment.

The experiments with the Gram-negative bacteria *Salmonella typhimurium* ATCC BAA-2162 and the antibiotic Ciprofloxacin showed similar results in the resistance of the bacteria against the antibiotic alone and in the combination of Ciprofloxacin with the highest concentration of the hybrid. The obtained results for *Salmonella typhimurium* are quite similar to those obtained for *Staphylococcus epidermidis* (Figure 6). The chart does not show a big difference in MBC between the independent action of the antibiotic and when it is combined with the highest concentration of the hybrid material SiO_2/15PVP (100 mg/mL). The MBC of the antibiotic acting alone against *Salmonella typhimurium* ATCC BAA-2162 was found to be 1.5 µg/mL, which is the same value as the effect of Ciprofloxacin in combination with SiO_2/15PVP. Interestingly, the data from the graph confirm the significant synergistic effect of the antibiotic in combination with SiO_2/15PVP, especially at their highest concentration of 100 mg/mL. As the minimum inhibitory concentration (MIC) of Ciprofloxacin in combination with the hybrid nanomaterials was 0.7 µg/mL, it is significantly lower than that of the antibiotic acting alone. The difference in the microbial quantity of the antibiotic alone and the combination of antimicrobial agents is almost four times more.

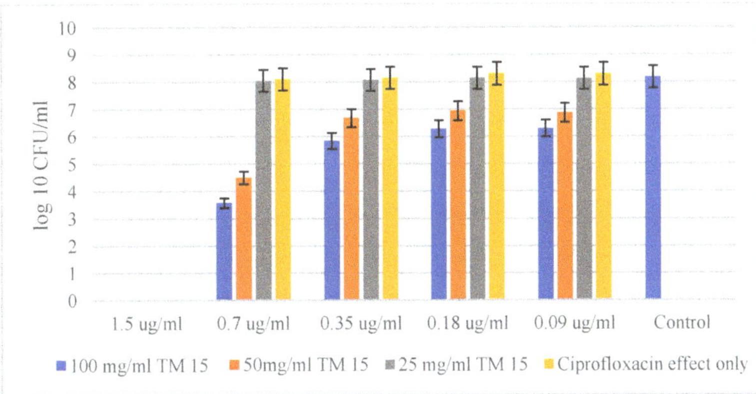

Figure 6. SiO_2/15PVP in combination with Ciprofloxacin against *Salmonella typhimurium* ATCC BAA-2162 at 24 h of the experiment.

2.5. Antifungal Effects

As mentioned above, SiO_2/15PVP hybrid nanomaterials were also tested in combination with the antifungal agent Nystatin against two eukaryotic cells. Of the two yeast cells tested, *Candida albicans* ATCC 18804 showed higher resistance to the concentration of antibiotics and the hybrid material. Here, the MBC of Nystatin alone was 12.5 µg/mL, while through a synergistic effect with the new hybrid material SiO_2/15PVP at a concentration of 100 mg/mL, the MBC decreased to 6.25 µg/mL, as shown in Figure 7.

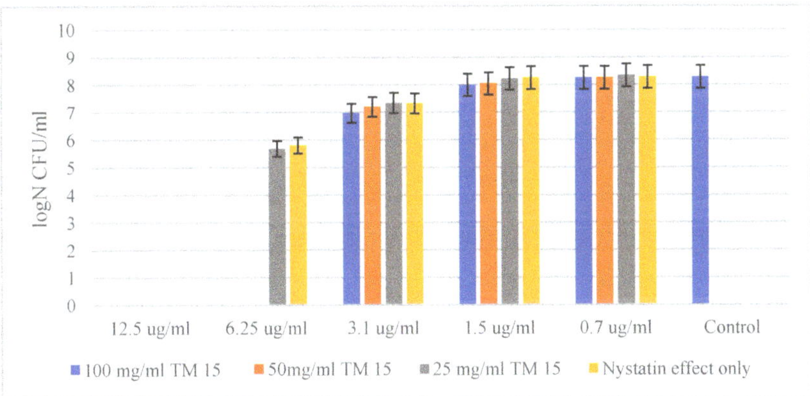

Figure 7. Effect of SiO_2/15PVP in combination with Nystatin against *Candida albicans* ATCC 18804 at 24 h of treatment.

The other yeast tested showed quite similar results in its suppression by Nystatin and the hybrid material. However, *Saccharomyces cerevisiae* CCY 21-6-3 turned out to be a much more sensitive microorganism. Figure 8 shows a similar MBC of Nystatin acting alone and in combination with 100 mg/mL of the hybrid material. The MBC of Nystatin against *Saccharomyces cerevisiae* was established to be 1.5 µg/mL, while for the combination of Nystatin and the highest concentration (100 mg/mL) of SiO_2/15PVP, the MIC was established to be 0.7 µg/mL. Microbial growth was inhibited by almost 10,000 times.

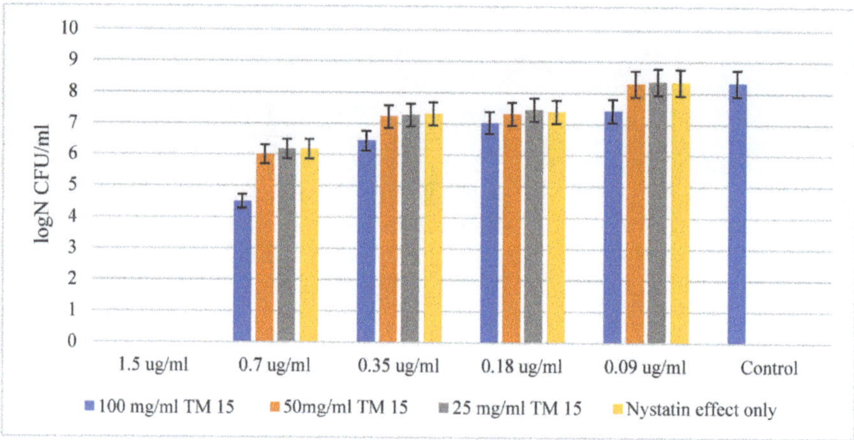

Figure 8. SiO_2/15PVP in combination with Nystatin against *Saccharomyces cerevisiae* CCY 21-6-3 at 24 h of the experiment.

2.6. Luminescence Tests for SiO_2/15PVP

The results of the chemiluminescence assay and the potential of the tested material to generate ROS and present antibacterial properties are presented in Figure 9.

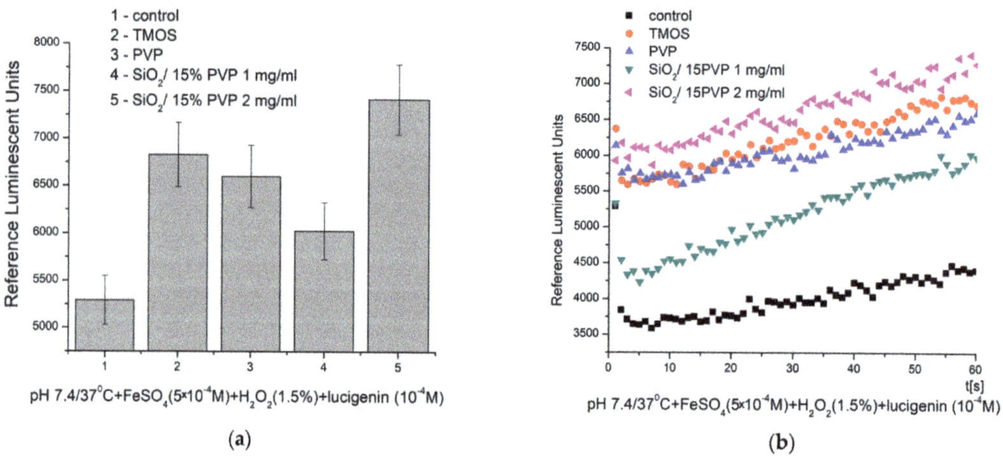

Figure 9. Chemiluminescence induced by ·OH and ·OOH radicals at pH 7.4/37 °C and the effect of the materials: (**a**) maximum effects and (**b**) Fenton's reaction over time ($p \leq 0.05$).

At pH 7.4/37 °C (physiological), the kinetic curves showed a gradual increase in the signal over time, indicating the generation of ROS in both the control and test reactions. The control reaction had the lowest signal, with a maximum of around 5300 RLU. In all other cases, the recorded signal was higher than the control, indicating the stimulated generation of ·OH and ·OOH radicals for the newly synthesized material and its precursors. The observed effects were as follows: TMOS—29%, PVP—25%, SiO_2/15PVP (1 mg/mL)—14%, and SiO_2/15PVP (2 mg/mL)—40%. These results demonstrate a significant, nearly threefold increase in the effect of the newly synthesized material at low, borderline application concentrations—2 in comparison to 1 mg/mL. The newly synthesized SiO_2/15PVP compound showed a significant pro-oxidant effect that persists over time. In comparison, our earlier results, described in [8], showed a similar prooxidant effect for the SiO_2/20PVP material in the same reaction and conditions at a significantly higher final concentration (25 mg/mL). The differences in the control reaction signal are due to the extreme sensitivity of the method, which is why controls are prepared for each evaluation.

2.7. Results on Daphnia Magna Toxicity Testing

Daphnia are extremely sensitive organisms in the aquatic environment, which allows conclusions to be drawn from the obtained results about the influence of SiO_2/15PVP on other aquatic organisms as well. Behavioral results of *Daphnia* showed 100% survival in the acute toxicity test at 0.001 and 0.0001 mg/mL concentration. At the highest concentration (0.01 mg/mL) at 24 h and 48 h, 97% of *Daphnia* survived. At the second concentration (0.005 mg/mL) up to the twenty-fourth day, survival was 100%, and at 48 h, 97% survival of *Daphnia* was observed. Compared to studies performed on the effect of PVP 20% on *D. magna* [7], the present results showed that PVP 15% was even better tolerated by *Daphnia*, close to 100%. Therefore, these concentrations of the substance used (PVP 15%) can be considered harmless and can be discharged for whatever reason into surface waters without harm to organisms (Figure 10).

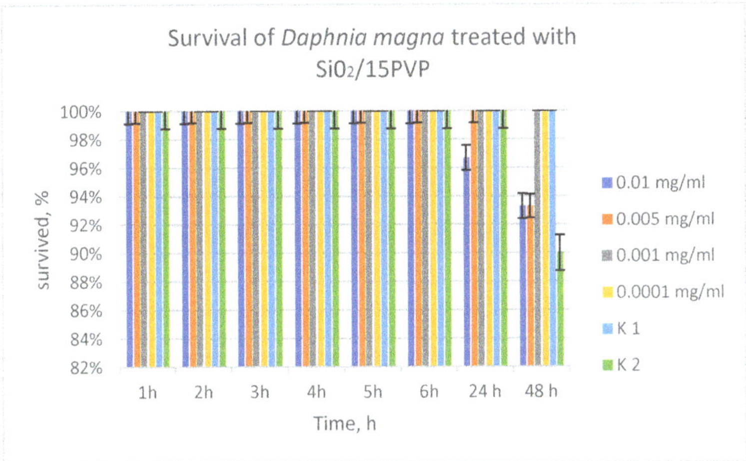

Figure 10. TM/15PVP without antibiotic influence on *Daphnia magna*.

2.8. Analysis of the Results

The above-presented experimental results confirm the trend that silica-containing materials possess a bacteriostatic effect [8,24,25]. These data correspond well with our previous data [8], but the present survey expands the investigations with other bacterial strains (*Staphylococcus epidermidis* ATCC 14990 and *Salmonella typhimurium* ATCC BAA-2162), as well as different fungi (*Candida albicans* ATCC 18804 and *Saccharomyces cerevisiae* CCY 21-6-3). Based on the obtained results, it could be concluded that all of the bacteria, both Gram-positive and Gram-negative, are sensitive to the presence of the hybrid material together with appropriate antibiotics. Once again, it has been demonstrated that the synergistic effect of the hybrid with antibiotics is specific to the tested bacteria. The antibacterial activity of the silica hybrid is evidently activated upon functionalization with relatively low concentrations of the antibiotics Ciprofloxacin, Vancomycin, and Nystatin. Obviously, the combination of the antibiotics and the SiO_2/15PVP material strongly affects the antibacterial properties of the synthesized material. The synergistic effects of Si/PVP with antibiotics against microorganisms have been described by several authors in the literature [14,26]. Generally, they can be explained by the formation of chelation bonds, which are considered one of the most important interactions as they increase the concentration of antimicrobial agents on the cell membrane. Perhaps, the synergistic effects of

Si/PVP with antibiotics against microorganisms could be explained by reactive oxygen species (ROS), which are formed in lower concentrations. Additional confirmation for the good antibacterial activity of the synthesized SiO_2/15PVP compound is the significant pro-oxidant effect that persists over time. These results provoke our scientific interest to continue these experiments, and future studies will focus on the investigation of the kind of prepared hybrid and the influence of the composition of the antibacterial properties. Such information could shed new light on the possibility of using these materials for several biomedical applications including drug delivery.

3. Experiment

3.1. *Materials and Preparation of Silica–Polyvinylpyrrolidone (SiO_2-PVP) Hybrids*

The gels were prepared using a combination of the alkoxide tetramethyl orthosilane (TMOS), Sigma Aldrich Chemical (St. Louis, MO, USA), and polyvynylpyrrolidon K25-(PVP), Fluka AG. (Buchs, Switzerland). The preparation procedure is described in detail below. The following precursors, tetramethyl orthosilicate {TMOS, $Si(OCH_3)_4$}, distilled H_2O, and 0.1 M HCl in a molar ratio of 4:1:1, were mixed and stirred together. PVP at 15 wt% was added with the aim of obtaining a hybrid material with improved properties. It was dissolved in ethanol and added dropwise to the stirred solution. Stirring was continued until the mixture became clear. Gelation occurred at room temperature after about 20 h. The investigated sample was denoted as SiO_2/15PVP.

3.2. *Methods for Characterization*

X-ray diffraction powder data were collected using a Philips PW1730/10 diffractometer using Ni-filtered CuKα radiation. The scanning rate for crystallinity was 11.2° 2θ/min. SEM images were obtained on a Hitachi S-4100 (Hitachi Ltd., Tokyo, Japan) microscope at an accelerating voltage of 25.0 kV. TEM images were captured using a Hitachi H-600A (Tokyo, Japan). A powdered hybrid specimen was dispersed in a mixture of ethanol and polysorbate 80 (1:1 v/v) by an ultrasonic generator (intensity of 250 W) for 3 min. One or two drops of the resulting dispersion were dropped on a 300-mesh copper grid (Tedpella, Inc., Redding, CA, USA) coated by carbon film and left to evaporate.

Atomic Force Microscopy (AFM) was used for the topographical contour mapping and roughness analysis of the model nanohybrid. Surfaces were mapped in the tapping mode (Intermittent Contact Mode) using NanoScope IIIa Dimension 3100 (Digital Instruments, Inc., Tokyo, Japan) with a prefabricated cantilever with a scan rate of 1.0 Herz; a nominal tip radius of 5–10 nm; a cantilever length of 125 μm; and a resonant frequency of ~200–400 Hz.

An "Evolution 300" UV–Vis diffused reflectance spectrophotometer, utilizing a magnesium oxide reflectance standard as a baseline, was employed to measure the optical absorption spectra of powdered samples within the wavelength range of 200 to 800 nm. The band gap energies (Eg) of the samples were calculated using Planck's equation, as already described elsewhere [27,28].

3.3. *Materials for Antimicrobial Activity Testing*

For the purpose of the present experiment, microorganisms provided by the National Bank for Industrial Microorganisms and Cell Cultures (NBIMCC, Sofia, Bulgaria) were used. Two types of opportunistic pathogenic bacteria were used—*Staphylococcus epidermidis* ATCC 14990 (Gram-positive) and *Salmonella typhimurium* ATCC BAA-2162, which is Gram-negative. In addition, two types of yeast were used—*Candida albicans* ATCC 18804, a commensal eukaryotic microorganism, and *Saccharomyces cerevisiae* CCY 21-6-3, a eukaryote widely used in the biotechnological industry. The SiO_2/15PVP hybrid nanomaterial was tested for antimicrobial efficacy alone and in combination with conventional antibiotics in liquid medium. Antibiotics were selected from five groups with a different mode of action, including Ciprofloxacin, an antibiotic belonging to the fluoroquinolone group. It is a broad-spectrum antibiotic and can be used to treat many bacterial infections. It exhibits activity against Gram-negative bacteria. Also, we used Vancomycin—a glycopeptide antibiotic used for complicated skin infections, bone and joint infections, and meningitis caused by MRSA. Its mechanism of action is aimed at the synthesis of the bacterial cell wall by binding to the basic building block. Nystatin was chosen as an antifungal agent as it is used to treat *Candida* skin infections and esophageal and vaginal candidiasis [29].

The minimum bactericidal concentration (MBC) of the sample alone and in combination with antibiotics against different microorganisms was examined using the broth microdilution method, as described by international protocols [30]. The control sample, used in all experiments, consisted of a pure culture of the strain in the exponential growth phase, with a McFarland OD of 0.5, and showed 10^8 bacterial colonies or 10^6 yeast colonies, demonstrating the typical growth of the target microorganism without the application of microbicidal agents—antibiotic and nanoparticles. In all experiments, reporting of results and enumeration of surviving cells occurred 24 h after treatment with nanoparticles, antibiotics, and a combination of both. Preparation of the hybrid nanomaterials included the following process: dispersions of nanocomposites in their respective concentrations (100 mg/mL, 50 mg/mL, and 25 mg/mL), by measuring the dry substance with analytical accuracy and dissolving it in distilled water. The respective antibiotics used for the purpose of the experiment were prepared in the same way.

3.4. *Methodology for Evaluation of the Minimum Inhibitory and Minimal Bactericidal Concentration (MIC) and Antibacterial Mode of Inhibition of SiO_2/15PVP*

The minimum inhibitory concentration (MIC) of the nanocomposites applied in combination with the antibiotic was determined by applying the microdilution method in a 96-well plate. Before starting the experiment, the respective antibiotic was diluted using a sterile 96-well plate. Next, 200 µL of the antibiotic at its initial concentration of 100 µg/mL was instilled into rows A1 and B1 of the 96-well plate. This was followed by a two-fold dilution of the antibiotic by taking 100 µL from the previous well and adding it to each subsequent well until reaching the lowest concentration in wells A11 and B11 of 0.09 µg/mL. When the antibiotic was ready for work, it was mixed with the hybrid material on a new plate: first, 50 µL of the antibiotic in its different concentrations was dripped into one well, and 50 µL of the hybrid material was added to it, again in the corresponding concentration. Finally, 100 µL of bacterial suspension was dripped onto the wells that contained the bactericidal agents. In column 12 of the 96-well plate, only the bacterial suspension was added, which serves as a control. The bacterial growth from the 96-well plate suspensions was monitored by preparing tenfold dilutions by dropping 100 µL onto solid agar and rubbing it until dry. The Petri dishes were incubated for 24 h, at the appropriate temperature, specific to each tested microorganism, and the resulting colonies from the respective dilution were counted. The amount of surviving treated microorganisms was determined by the following formula: CFU/mL (number of colonies $\times$ dilution factor)/inoculum sample volume. The minimum bactericidal concentration (MBC) demonstrates the lowest level of antimicrobial agent that results in microbial death. The MBC is identified by determining the lowest concentration of antibacterial agent that reduces the viability of the initial bacterial inoculum by $\geq$99.9% [31]. Experiments were performed in three independent replicates, and values from each replicate were averaged.

3.5. *Chemiluminescence Assay*

3.5.1. Materials

We purchased the following compounds at high purity: iron sulfate (p.a.) (Merck, Darmstadt, Germany), hydrogen peroxide (30%) (Merck, Darmstadt, Germany), lucigenin (bis-*N*-methylacridinium nitrate) (p.a.) (Sigma-Aldrich, St. Louis, MO, USA), dimethyl sulfoxide (p.a.) (DMSO, Sigma-Aldrich, St. Louis, MO, USA), and buffer pH 7.4 (Sigma-Aldrich, St. Louis, MO, USA). All chemicals were used as purchased.

3.5.2. Method

The chemical probe that we applied for signal amplification is lucigenin. Thus, reliable, comparable differences were achieved.

Fenton's model system generates ·OOH and ·OH radicals [12]. The control samples do not contain the tested material (which was applied in final concentrations of 1 or 2 mg/mL). The reactions were monitored for 3 min, every 3 s, measured in triplicate, and presented as average and standard deviation values. All tested materials were sonicated for 30 min prior to testing to ensure good dispersion.

Fenton's system

(1) $Fe^{2+} + H_2O_2 \rightarrow Fe^{3+} + \cdot OH + {}^{-}OH$

(2) $Fe^{3+} + H_2O_2 \rightarrow Fe^{2+} + \cdot OOH + H^+$

3.5.3. Statistics

All experiments were performed using LUMIstar Omega (BMG Labtech GmbH, Ortenberg, Germany, 2020) in triple-reproducible measurements; statistical analysis was performed with OriginPro 8 and Microsoft Office Excel 2010.

3.6. *Daphnia Magna Toxicity Testing*

An acute toxicity test of the resulting SiO_2/15PVP hybrid was conducted on *Daphnia magna*, with the aim of predicting the maximum permissible concentration in the environment without harm to aquatic organisms. The experiments used 4 concentrations of the substance (0.01 mg/mL; 0.005 mg/mL; 0.001 mg/mL; 0.0001 mg/mL) with 3 replicates of each concentration and 2 controls. The results were detected at 48 h and processed statistically with the Excel program.

4. Conclusions

The synthesis of hybrid nanocomposites containing TMOS and poly (vinylpyrrolidone) was carried out via the sol–gel method at room temperature. The morphological structure of the final product was determined by the incorporated PVP. The observed UV–Vis analysis results showed that the obtained hybrid exhibited absorption in the ultraviolet range.

The as-prepared SiO_2/15PVP hybrid demonstrated a bacteriostatic effect, but its efficacy decreased with decreasing concentration. This was evidenced by the higher number of surviving microorganisms at lower concentrations of SiO_2/15PVP. The antibiotics Vancomycin and Ciprofloxacin were significantly more effective when applied in combination with SiO_2/15PVP in inhibiting the growth of *Staphylococcus epidermidis* ATCC 14990 and *Salmonella typhimurium* ATCC BAA-2162, at all antibiotic concentrations tested. The action of Nystatin was significantly enhanced when it was administered in combination with the hybrid materials, especially at a concentration of 100 mg/mL of SiO_2/15PVP.

The registered and confirmed prooxidant activity of SiO_2/15PVP shows possible applications at very low concentrations. A significant, nearly threefold increase in the effect of the newly synthesized material at 2 mg/mL in comparison to 1 mg/mL was observed.

Daphnia magna was more sensitive than the bacteria to the tested nanoparticles. The harmless concentration to the environment is 0.01 mg/mL according to the conducted experiments.

The obtained results demonstrate the possibility of various clinical implementations of the new synthesized hybrid material.

Author Contributions: Conceptualization, Y.K. and A.B.-N.; methodology, Y.K. and A.B.-N.; sample characterization, Y.K. and A.B.-N.; antibacterial properties, I.I. and L.Y.; chemiluminescence assay, E.P.; toxicity test with Daphnia magna, E.N.; writing, A.B.-N., Y.K., L.Y., I.I. and E.P. All authors have read and agreed to the published version of the manuscript.

Funding: The team from the Faculty of Biology, Sofia University, is grateful for the financial support provided by the European Union Next Generation EU through the National Recovery and Resilience Plan of the Republic of Bulgaria, project № BG-RRP-2.004-0008-C01 contract 76-123-661, and the scientific fund of Sofia University project 80-10-176/18.04.2024.

Institutional Review Board Statement: Not applicable.

Informed Consent Statement: Not applicable.

Data Availability Statement: Data are contained within the article.

Acknowledgments: The author A. Bachvarova-Nedelcheva is thankful to the "National Center of Excellence Mechatronics and Clean Tecnologies" for the experimental work supported by the European Regional Development Fund under the "Research Innovation and Digitization for Smart Transformation" program 2021–2027. Research equipment of distributed research infrastructure INFRAMAT, supported by the Bulgarian Ministry of Education and Science was used. Elitsa Pavlova is grateful to the project: CLEAN TECHNOLOGIES FOR SUSTAINABLE ENVIRONMENT—WATERS, WASTE, ENERGY FOR CIRCULAR ECONOMY, Ministry of Education and Science, Bulgaria, contract number: BG05M2OP001-1.002-0019. The team from the Faculty of Biology, Sofia University, is grateful to the support provided by the European Union Next Generation EU through the National Recovery and Resilience Plan of the Republic of Bulgaria, project № BG-RRP-2.004-0008-C01 contract 76-123-661, and the scientific fund of Sofia University project 80-10-176/18.04.2024.

Conflicts of Interest: The authors declare no conflicts of interest.

References

1. Toki, M.; Chow, T.Y.; Ohnaka, T.; Samura, H.; Saegusa, T. Structure of poly(vinylpyrrolidone)-silica hybrid. *Polym. Bull.* **1992**, *29*, 653–660. [CrossRef]
2. Urquijo-Ventura, M.S.; Syamala Rao, M.G.; Meraz-Davila, S.; Torres- Ochoa, J.A.; Quevedo-Lopez, M.A.; Ramirez-Bon, R. PVP-SiO_2 and PVP-TiO_2 hybrid films for dielectric gate applications in CdS-based thin film transistors. *Polymer* **2020**, *191*, 122261. [CrossRef]
3. Albalawi, F.; Hussein, M.Z.; Fakurazi, S.; Masarudin, M.J. The Challenges and Opportunities for Nanomedicines. *Int. J. Nanomed.* **2021**, *16*, 161–184. [CrossRef] [PubMed]
4. Rejab, M.R.; Hamdan, M.H.; Quanjin, M.; Siregar, J.P.; Bachtiar, D.; Muchlis, Y. Historical Development of Hybrid Materials. *Mat. Sci. Mat. Eng.* **2020**, *4*, 445–455.
5. Salam, M.A.; Al-Amin, M.Y.; Salam, M.T.; Pawar, J.S.; Akhter, N.; Rabaan, A.A.; Alqumber, M.A.A. Antimicrobial Resistance: A Growing Serious Threat for Global Public Health. *Healthcare* **2023**, *11*, 1946. [CrossRef]
6. Meng, H.; Mai, W.X.; Zhang, H.; Xue, M.; Xia, T.; Lin, S.; Wang, X.; Zhao, Y.; Ji, Z.; Zink, J.I.; et al. Co-delivery of an Optimal Drug/siRNA 82 Combination Using Mesoporous Silica Nanoparticle to Overcome Drug Resistance in Breast Cancer In Vitro and In Vivo. *ACS Nano* **2013**, *7*, 994–1005. [CrossRef]
7. Ilhan-Ayisigi, E.; Yesil-Celiktas, O. Silica-based organic-inorganic hybrid nanoparticles and nanoconjugates for improved anticancer drug delivery. *Eng. Life Sci.* **2018**, *18*, 882–892. [CrossRef]
8. Bachvarova-Nedelcheva, A.; Kostova, Y.; Yordanova, L.; Nenova, E.; Shestakova, P.; Ivanova, I.; Pavlova, E. Sol-Gel Synthesis of Silica-Poly (Vinylpyrrolidone) Hybrids with Prooxidant Activity and Antibacterial Properties. *Molecules* **2024**, *29*, 2675. [CrossRef]
9. Abebe, B.; Zereffa, E.A.; Tadesse, A.; Murthy, H.A. A Review on Enhancing the Antibacterial Activity of ZnO: Mechanisms and Microscopic Investigation. *Nanoscale Res. Lett.* **2020**, *15*, 190. [CrossRef]
10. Pavlova, E.L.; Toshkovska, R.D.; Doncheva, T.E.; Ivanova, I.A. Prooxidant and antimicrobic effects of iron and titanium oxide nanoparticles and thalicarpine. *Arch. Microbiol.* **2020**, *202*, 1873–1880. [CrossRef]
11. Pavlova, E.L.; Ivanova, I.A.; Staneva, A.D.; Kostadinova, A.S.; Kichukova, D.G.; Yocheva, L.D. Prooxidant, antioxidant and biological activity of nanocomposites of reduced graphene oxide, silver, copper and their combinations. *Chem. Pap.* **2022**, *76*, 6789–6800. [CrossRef]
12. Faulkner, K.; Fridovich, I. Luminol and lucigenin as detectors for O_2. *Free Radic. Biol. Med.* **1993**, *15*, 447–451. [CrossRef] [PubMed]
13. Allahverdiyev, A.M.; Kon, K.V.; Abamor, E.S.; Bagirova, M.; Rafailovich, M. Coping with antibiotic resistance: Combining nanoparticles with antibiotics and other antimicrobial agents. *Expert Rev. Anti Infect. Ther.* **2011**, *9*, 1035–1052. [CrossRef] [PubMed]
14. Ribeiro, A.I.; Dias, A.M.; Zille, A. Synergistic Effects Between Metal Nanoparticles and Commercial Antimicrobial Agents: A Review. *ACS Appl. Nano Mater.* **2022**, *5*, 3030–3064. [CrossRef]
15. Ziesmer, J.; Larsson, J.V.; Sotiriou, G.A. Hybrid microneedle arrays for antibiotic and near-IR photothermal synergistic antimicrobial effect against Methicillin-Resistant *Staphylococcus aureus*. *Chem. Eng. J.* **2023**, *462*, 142127. [CrossRef] [PubMed]

16. Abdullah, J.; Atif, M.; Khalid, S.; Metwally, K.; Yahya, G.; Moisa, M.; Cavalu, D.S. Recent Advances in the Development of Metal/Metal Oxide Nanoparticle and Antibiotic Conjugates (MNP–Antibiotics) to Address Antibiotic Resistance: Review and Perspective. *Int. J. Mol. Sci.* **2024**, *25*, 8915. [CrossRef]
17. Xu, X.; Xu, L.; Yuan, G.; Wang, Y.; Qu, Y.; Zhou, M. Synergistic combination of two antimicrobial agents closing each other's mutant selection windows to prevent antimicrobial resistance. *Sci. Rep.* **2018**, *8*, 7237. [CrossRef]
18. Timin, A.S.; Solomonov, A.V.; Musabirov, I.I.; Sergeev, S.N.; Ivanov, S.P.; Rumyantsev, E.V.; Goncharenko, A. Immobilization of Bovine Serum Albumin onto Porous Poly(vinylpyrrolidone)-Modified Silicas. *Ind. Eng. Chem. Res.* **2014**, *53*, 13699–13710. [CrossRef]
19. Tański, T.; Matysiak, W.; Krzemiński, Ł.; Jarka, P.; Gołombek, K. Optical properties of thin fibrous PVP/SiO_2 composite mats prepared via the sol-gel and electrospinning methods. *Appl. Surf. Sci.* **2017**, *424 Pt 2*, 184–189. [CrossRef]
20. Xu, X.; Wang, M.; Pei, Y.; Ai, C.; Yuan, L. SiO_2@Ag/AgCl: A low-cost and highly efficient plasmonic photocatalyst for degrading rhodamine B under visible light irradiation. *RSC Adv.* **2014**, *4*, 64747–64755. [CrossRef]
21. Weinberg, Z.A.; Rubloff, G.W.; Bassous, E. Transmission, photoconductivity, and the experimental band gap of thermally grown SiO_2 films. *Phys. Rev. B* **1979**, *19*, 3107. [CrossRef]
22. Aleksandrova, M.; Petkov, V.; Blaskov, V.; Korzhov, V.; Kiiko, V.; Zheltyakova, I. Synthesis and Properties of Sol-gel Coatings Deposited on Hybrid Materials. *C. R. Acad. Bulg. Sci.* **2023**, *76*, 1269–1276. [CrossRef]
23. Purcar, V.; Răditoiu, V.; Răditoiu, A.; Manea, R.; Raduly, F.M.; Ispas, G.C.; Frone, A.N.; Nicolae, C.A.; Nicolae, R.A.; Anastasescu, M.; et al. Preparation and Characterization of Some Sol-Gel Modified Silica Coatings Deposited on Polyvinyl Chloride (PVC) Substrates. *Coatings* **2021**, *11*, 11. [CrossRef]
24. Blanco, I.; Latteri, A.; Cicala, G.; D'Angelo, A.; Viola, V.; Arconati, V.; Catauro, M. Antibacterial and Chemical Characterization of Silica-Quercetin-PEG Hybrid Materials Synthesized by Sol–Gel Route. *Molecules* **2022**, *27*, 979. [CrossRef]
25. Varshney, S.; Nigam, A.; Pawar, S.J.; Mishra, N. Structural, optical, cytotoxic, and anti-microbial properties of amorphous silica nanoparticles synthesised via hybrid method for biomedical applications. *Mater. Technol.* **2021**, *37*, 1504–1515. [CrossRef]
26. Dizaj, S.M.; Lotfipour, F.; Barzegar-Jalali, M.; Zarrintan, M.H.; Adibkia, K. Antimicrobial activity of the metals and metal oxide nanoparticles. *Mater. Sci. Eng. C* **2014**, *44*, 278–284. [CrossRef]
27. Jameel, M.H.; Hamzah, M.Q.; Bin Agam, M.A.; Alper, M. Synthesis and Characterizations of Co-Doped TiO_2 Nanoparticles Via Co-Precipitation Method. *Solid State Technol.* **2020**, *63*, 267–277.
28. Stoyanova, A.; Hitkova, H.; Ivanova, N.; Bachvarova-Nedelcheva, A.; Iordanova, R.; Sredkova, M. Photocatalytic and antibacterial activity of Fe-doped TiO_2 nanoparticles prepared by nonhydrolytic sol-gel method. *Bulg. Chem. Commun.* **2013**, *45*, 497–504.
29. Lyu, X.; Zhao, C.; Yan, Z.M.; Hua, H. Efficacy of nystatin for the treatment of oral candidiasis: A systematic review and meta-analysis. *Drug Des. Dev. Ther.* **2016**, *10*, 1161–1171. [CrossRef]
30. Mogana, R.; Adhikari, A.; Tzar, M.N.; Ramliza, R.; Wiart, C. Antibacterial activities of the extracts, fractions and isolated compounds from Canarium patentinervium Miq. against bacterial clinical isolates. *BMC Complement. Med. Ther.* **2020**, *20*, 1–11. [CrossRef]
31. Levison, M.E. Pharmacodynamics of antimicrobial drugs. *Infect. Dis. Clin. N. Am.* **2004**, *18*, 451–465. [CrossRef] [PubMed]

Disclaimer/Publisher's Note: The statements, opinions and data contained in all publications are solely those of the individual author(s) and contributor(s) and not of MDPI and/or the editor(s). MDPI and/or the editor(s) disclaim responsibility for any injury to people or property resulting from any ideas, methods, instructions or products referred to in the content.

molecules

MDPI

Communication

Novel Approach for the Preparation of a Highly Hydrophobic Coating Material Exhibiting Self-Healing Properties

Uwe Holzdörfer [1], Wael Ali [1,2], Eckhard Schollmeyer [1], Jochen S. Gutmann [1,2,*], Thomas Mayer-Gall [1,2,*] and Torsten Textor [3,*]

1 Deutsches Textilforschungszentrum Nord-West gGmbH, Adlerstr. 1, 47798 Krefeld, Germany
2 Physikalische Chemie, Center for Nanointegration Duisburg-Essen, Universität Duisburg-Essen, Universitätsstraße 2, 45117 Essen, Germany
3 TEXOVERSUM School of Textiles, Reutlingen University, Alteburgstr. 150, 72762 Reutlingen, Germany
* Correspondence: jochen.gutmann@uni-due.de (J.S.G.); mayer-gall@dtnw.de (T.M.-G.); torsten.textor@reutlingen-university.de (T.T.)

Abstract: A concept to prepare a highly hydrophobic composite with self-healing properties has been designed and verified. The new material is based on a composite of a crystalline hydrophobic fluoro wax, synthesized from montan waxes and perfluoroethylene alcohols, combined with spherical silica nanoparticles equipped with a hydrophobic shell. Highly repellent layers were prepared using this combination of a hydrophobic crystalline wax and silica nanoparticles. The novel aspect of our concept was to prepare a ladder-like structure of the hydrophobic shell allowing the inclusion of a certain share of wax molecules. Wax molecules trapped in the hydrophobic structure during mixing are hindered from crystallizing; therefore, these molecules maintain a higher mobility compared to crystallized molecules. When a thin layer of the composite material is mechanically damaged, the mobile wax molecules can migrate and heal the defects to a certain extent. The general preparation of the composite is described and XRD analysis demonstrated that a certain share of wax molecules in the composite are hindered to crystallize. Furthermore, we show that the resulting material can recovery its repellent properties after surface damage.

Keywords: self-healing; hydrophobicity; sol-gel; silica nanoparticle

1. Introduction

In the late 1940s, Wenzel [1–3] as well as Cassie and Baxter [4,5] explained the influence of the topography of a surface with a given chemical composition on repellence. The Cassie–Baxter model predicts an increase in apparent contact angle with the increasing roughness of an inherently hydrophobic surface. Optimal repellence can be achieved with structures featuring both micro- and nanometer scale. Since that time, scientists have been striving to develop hydrophobic surfaces with minimal surface energy combined with specific surface structures to create super-repellent surfaces [6–12]. The primary challenge is to prepare a durable hydrophobic surface structure in the micrometer range. Maximum repellence will be achieved when an additional structure in the nanometer range is also incorporated.

Various approaches have been investigated leading to more or less successful results. A significant issue with structured hydrophobic surfaces is their comparatively low mechanical resistance. During use, the super-repellent surfaces can become abraded, resulting in flat hydrophobic surfaces at best, and hydrophilic surfaces at worst if the abrasion not only flattens the structure but also removes the hydrophobic coating completely [13]. Different plant waxes are suitable to recover the wear [14]. As a biomimetic approach, nanotechnology has emerged as a promising approach to develop micro- and possibly nanostructured hydrophobic surfaces with improved wear resistance [15–18]. The nanoparticles employed are expected to improve the abrasion resistance and can also serve as shape-forming templates to achieve a desired topography.

As most approaches employing nanoparticles for the preparation of super-repellent surfaces are focused on an improvement in the wear-resistance, the authors' superior aim was to develop novel composites capable of recovering the super-repellent properties if damaged by abrasion. The basic approach was to prepare a composite based on inorganic nanoparticles with surfaces modified by long-chained highly fluorinated alkoxysilanes, combined with a new extremely hydrophobic highly crystalline wax system [15–18].

The inclusion of a certain share of wax into hydrophobic structures on the particle surface will prevents the crystallization of these wax molecules and guarantee a degree of mobility (compared to the crystalline share of wax) of the hydrophobic material in case of a physical damage to the super-repellent surface. This mobility allows for a certain recovery of the damaged layer. Thus, the combination of hydrophobic nanoparticles and hydrophobic wax guarantees highly or even super-repellent surfaces. Self-healing can be achieved if the mobility of the wax molecules, hindered from crystallizing by inclusion into the hydrophobic structure, is sufficient to close—"repair"—damaged areas.

This approach is also applied by different other authors. Different waxes [19–21], bio-based or fluor-containing systems, have been used to obtain hydrophobic and self-repairing coatings. A fluorine-free approach demonstrates that self-healing and hydro- or superhydrophobicity function depends on the particle size and the particle-to-wax ratio [22]. Other approaches use self-lubricating slippery surfaces, which store a hydrophobic solvent in a nanoporous surface. The stored solvent prevents the surface from damaging [23–27]. By using dynamic covalent bond chemistry or vitrimers also, self-healing hydrophobic surfaces are possible [28–30]. All self-healing approaches have in common that these processes take a certain time or need activation energy in the form of heat or light.

Self-healing hydrophobic surfaces are interesting not only for the mentioned example of a high-class mountaineering cord. In applications where material can suffer abrasive damage like outdoor clothes, roofing membranes, personal protective equipment like work wear or tents, these properties are highly desired.

The hydrophobic highly crystalline wax system used in this study was synthesized based on the so-called montan waxes. These montan waxes [31–35] are byproducts of brown coal mining that are separated from coal by extraction. The raw product is basically a mixture of long-chained fatty acid esters and alcohols (C_{22}–C_{34}) and has been characterized in a very detailed manner within the framework of the overarching aim of the underlying research project. Following our approach, the montan waxes have to be purified, the esters have to be saponified and the long-chained alcohols have to be oxidized to yield a mixture of the corresponding long-chained, unbranched fatty acids with a high purity. Alternatively, a commercially available material (Luwax® S, BASF) has been used. An extremely hydrophobic crystalline wax was synthesized by esterification of the acids with perfluoroalkylethyl alcohols (C_8–C_{12}). While a more detailed description of the wax and its preparation is not necessary here, it is important to note that the novel approach for preparing self-healing material relies on using a hydrophobic wax based on linear molecules with a strong tendency to crystallize.

2. Results and Discussion

2.1. Preparation of Hydrophobic Nano Particles

One of the ideas of our approach was to employ nanoparticles modified with hydrophobic surface groups that allow and facilitate the inclusion of wax molecules, thus preventing their crystallization. For the first investigations, Stöber particles [36] were employed and modified with a long-chained, highly fluorinated alkoxysilane. Commercially available silanes typically have relatively short fluorocarbon chains (maximum C_6). However, lower surface energies as well as a more effective inclusion of wax molecules are expected using hydrophobic silanes with longer chains. To this end, a silane with a C10-

fluorocarbon chain was synthesized by reflux heating 3-isocyanatopropyltriethoxysilane and α,α,ω-trihydroperfluoroundecanol in the presence of triethanolamine and tin(II) octoate at a temperature of 80 °C under a nitrogen atmosphere (Figure 1A).

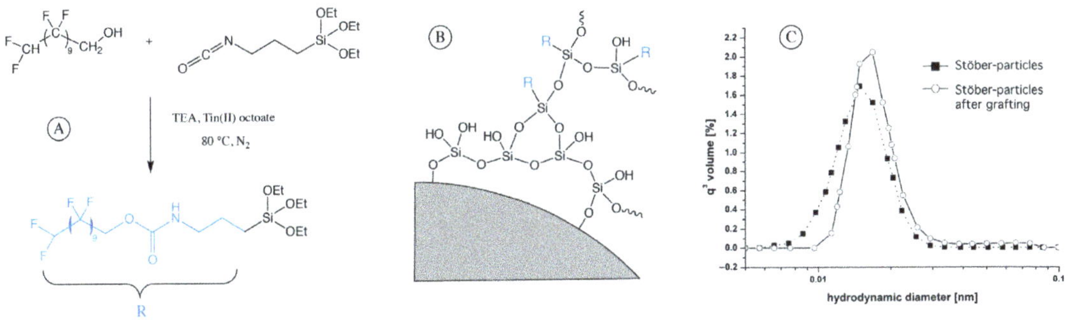

Figure 1. (**A**) Reaction scheme for the synthesis of the fluorinated alkoxy silane. (**B**) Modification of the silica particle surface. (**C**) DLS measurement of Stöber particles before and after grafting of fluorinated silane.

Silica particles in methyl-2-pentanone and the synthesized alkoxysilane were mixed prior to the addition of hydrochloric acid, which served as a catalyst to initiate the grafting reactions (Figure 1B). The silanol groups on the surface of the silica particles were quantified following the method described by Gorlov [37], which allowed to set a one-to-one ratio between silanol groups and the synthesized alkoxysilane. The grafting process was carried out by refluxing the mixture for 48 h. Subsequentially, small amounts of precipitates were filtered off before the addition of hexamethyldisilazane, which was used to end-cap any remaining silanol groups, thus enhancing the ageing stability of the sol. Ideally, the grafting reaction would result in a monomeric coverage. However, the formation of "ladder structures" is anticipated to be more realistic, although it has not been proven here. For silica particles of an initial size of 15.4 nm, dynamic light scattering (DLS) measurements indicated an increase in hydrodynamic diameter of the particle by approximately 1.6 nm (Figure 1C).

2.2. *Preparation of the Highly Repellent Wax*

The wax used to prepare highly repellent surfaces was synthesized from montan waxes, which are naturally occurring mixtures of C_{22}–C_{34} fatty acids and alcohols. The linear structure of these acids and alcohols allows the formation of crystalline waxes. Commercially available purified montan waxes, which contain only the fatty acids, provide a certain repellence but only for more polar liquids. To enhance repellent properties, fluorocarbon groups were introduced via esterification. Several esterification processes were tested, but satisfactory yields were only achieved when catalyzing the reaction with tetrabutyl titanate [35], resulting in ~97% yield. Following the distillation of the byproduct water and cooling to ambient temperature, a pale waxy material was obtained. As supposed, XRD measurements confirmed a strong tendency of the resulting ester to crystallize, as evidenced by the prominent peaks observed in Figure 2.

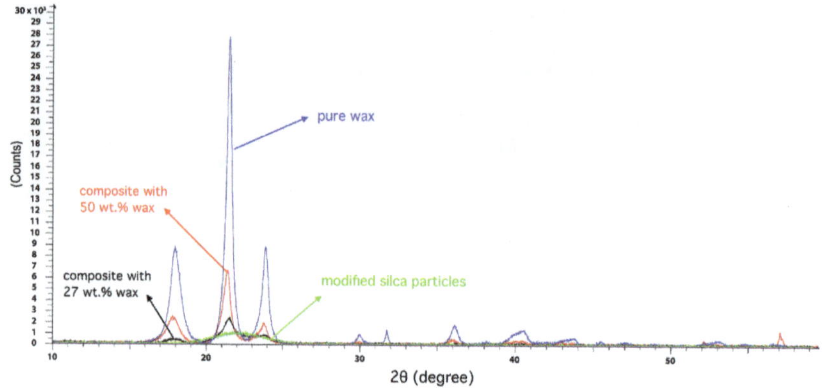

Figure 2. XRD spectra of the pure wax, the modified silica nanoparticles and two composites containing different amounts of wax. (broad, flat spectrum: silica nanoparticles only, the other spectra represent measurements for composites with 27, 50 and 100 wt.% wax—the higher the intensity the more the share of wax).

2.3. Preparation and Application of the Composite

The composite material was prepared in a simple manner by mixing the fluorinated wax with the sol containing the modified silica nanoparticles using methyl-2-pentanone as a solvent. The desired ratios of wax to sol were achieved by stirring the mixture while heating it to 60 °C, yielding a clear solution. This solution was then applied to glass slides (as model substrate) as well as textile substrates through casting and padding, respectively. The samples were subsequently dried to remove the solvent. The SEM images of two composites prepared with silica particles of different sizes are shown in Figure 3. The micrographs reveal that the composite with very small silica particles (~15 nm) displays a topography indicative of a strong tendency for the material to crystallize. In contrast, composites prepared with larger particles (~286 nm) exhibit a microstructured surface.

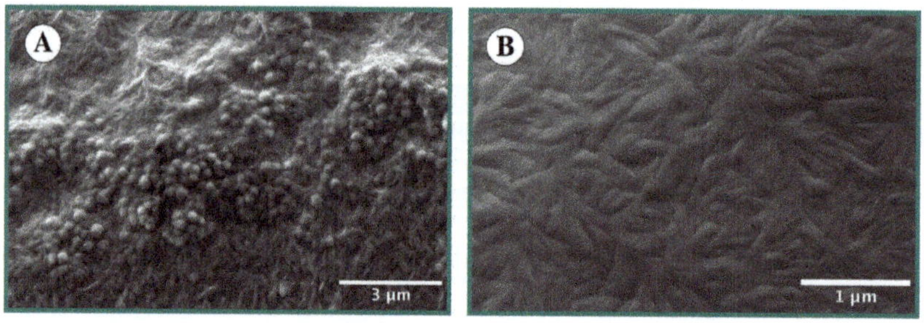

Figure 3. SEM micrographs of a composite prepared from the fluorinated wax and modified silica particles of approx. 15 nm (**B**) and corresponding particles of 286 nm (**A**).

2.4. Investigation of the Expected Inclusion of Wax Molecules

XRD measurements were carried out with selected samples to investigate the potential intercalation of wax molecules between the hydrophobic surface groups of the silica particles. XRD spectra were recorded of the silica particles (after surface modification with fluorocarbon groups), the pure wax (after esterification reaction with fluorocarbon groups) and two composites with different shares of nanoparticles. The corresponding XRD spectra are shown in Figure 2.

As expected, the amorphous silica particles show no diffraction peaks, while the fluorinated wax shows three strong signals. Adding certain amounts of nanoparticles to the wax resulted in a decrease in signal intensity, due to the reduction in the proportion of crystalline material forming the composite. A more detailed analysis of the integrated signal intensities is shown in Figure 4. The data reveal that the integrals of the signal for the nanoparticle–wax composites, for a given amount of wax, are smaller than those of the pure wax. This suggests that the wax crystallizes to a lower degree when the composite is prepared, which aligns well with our expectations.

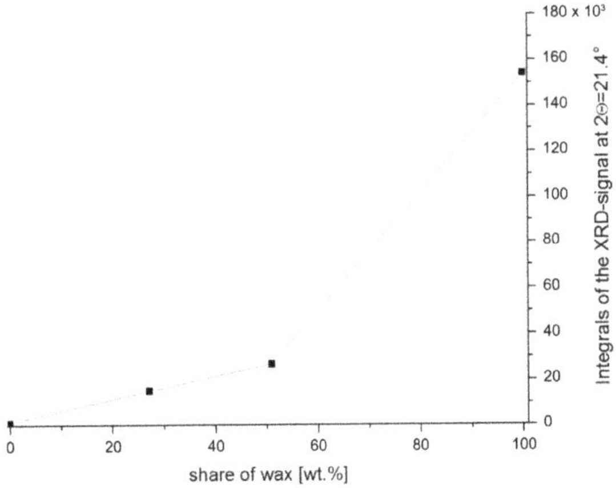

Figure 4. Graphical analysis of the integral of diffraction peaks at $2\theta = 21.4°$ as a function of the wax proportion in the investigated materials (solid line is guide to eyes).

The aim of this study was to develop a hydrophobic material. Therefore, the surface energy and contact angle of the resulting materials are of fundamental importance concerning the desired properties. Surface energy was calculated following the approach of Owens and Wendt [38]. Contact angles were measured for water (143.0° ± 0.1) and for diiodomethane (132.4.0° ± 0.2) on coatings applied to glass slides. The SEM of the coated surface (see Figure 3A) reveal that the surface is not flat, which explains the discrepancies in the contact angle measurements, as the Owens and Wendt method requires measurements on flat surfaces, which is not possible for the composite investigated here. Nevertheless, the results demonstrate that the composite exhibits low surface energy and an increased repellence, attributable to both surface energy and the surface structure. The roughness of the coatings was investigated by atomic force microscopy (AFM) measurements, with the roughness (R_a) calculated to be 200 nm.

An important aspect of this study was the self-healing ability of the highly hydrophobic layer. To evaluate this property, several technical products were coated, and the surfaces were damaged via plasma etching. The commercial product finished with two composites was a high-class mountaineering cord. These cords require high repellence, as a water-soaked rope significantly increases the risk of tearing when the climber falls dramatically. The repellent coating must withstand damage from friction during use, as defects ingress via capillary forces. Ropes coated with the hydrophobic composite were subjected to plasma etching, and the contact angle after recovery was monitored (see Table 1 and Figure 5). The contact angle of about 119° before etching, was reduced to 57° due to the plasma process. The recovery of the contact angle within 168 h was less than expected, reaching only a value of 72.5°. However, additional tempering at 40 °C resulted in a notable recovery and a contact angle of at least 94.2° was achieved. This indicates that the mobility of the wax molecules is insufficient for recovery within a relevant timescale under ambient conditions.

Table 1. The contact angle within 1 to 168 h after plasma etching of the repellent composite surfaces. The evaluation was performed for the combination of the modified particles with two different waxes. The change in contact angle before and after various durations following surface etching is given.

	As Prepared	1 h	16 h	38 h	168 h	168 h + Storage at 40 °C
original (high crystallin)	119	57	62	73	70	94
alternative wax (low crystallin)	105	68	100	100	99	100

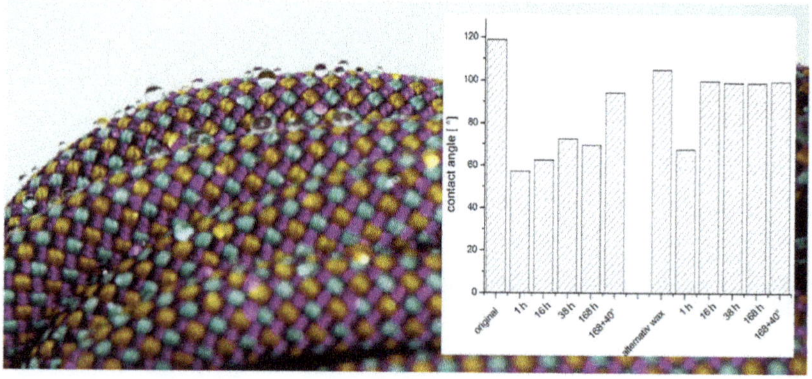

Figure 5. Water droplets on a rope and their contact angle measured after plasma etching and subsequent coating recovery over different time durations. A comparison was made between the original system and an alternative wax.

To validate the general concept, an alternative composite was prepared by replacing the wax with a less crystalline one, prepared from 2-butyloctanoic acid and the perfluoroethylene alcohol. The branched alkyl rest is expected to reduce the crystallinity of the resulting wax. The contact angle achieved by the alternative composite was approximately 104.8°, which is lower than that of the previously tested composite tested (see Table 1). Plasma etching again resulted in a significantly decreased contact angle for one hour in recovery. However, the alternative composite showed a relatively fast and nearly complete recovery, achieving a stable contact angle (~100°) from 16 h onward. While plasma etching provides valuable information on the recovery ability of the treated surfaces, it is important to note that etching not only causes physical damaging but also leads to oxidation, increasing the hydrophilicity of the wax material at the surface. Finished ropes were also subjected to realistic operational tests, e.g., a treatment with a so-called rope-tester, which simulates wear damage. The product showed comparable performance in this specific test. The self-healing effect works by the inclusion of a certain fraction of wax into hydrophobic structures on nanoparticle surfaces. This will prevent the crystallization of these wax molecules, whereby the degree of mobility of the hydrophobic material is increased. In case of a physical damage event, this mobility allows a recovery of the damaged layer, which is observed by regaining a higher contact angle after a certain time after the damaging of the surface.

From the authors' point of view, the results support the basic concept for a self-healing repellent material that was followed here. Future investigation will be needed to describe the alternative composite in more detail as presented for the one based on montan wax. Celik et al. demonstrated a similar but fluorine-free and bio-based approach based on carnauba wax. In this case, the right particle size and nanoparticle-to-wax ratio had to be selected to obtain a self-healing and hydrophobic surface [22].

Besides being a scientifically interesting system, it can be stated that the ropes finished with the described composites outperform commercially available products.

3. Materials and Methods

3.1. Materials

3-Isocyanatopropyltriethoxysilane (GE-Crompton, >95%) was purified by vacuum distillation to remove impurities. α,α,ω-trihydroperfluoroundecanol supplied by Halogen Russia (Perm) (92.0%) was also purified via vacuum distillation before use. The following substances were used without further purification: ammonia 25% p.a. (Fluka, Buchs, Switzerland), 2-Butyl-1-octanol (Isofol 12, >97% Sasol, Hamburg, Germany), ethanol (>99.9%; H_2O: <0.05% Alcosuisse, Rüti bei Büren, Switzerland), hexamethyldisilazane (HMDS, >98%, Fluka), hexane (>99.0%, Fluka), tetraethoxysilane (TEOS, >99.9%), tin(II), octoate (Metatin S 28, Rohm & Haas, Acima, Buchs, Switzerland), MIBK (4-Methyl-2-pentanone, >98%), Fluka and Luwax S (BASF, San Bruno, CA, USA)).

3.2. Analytic

Infrared spectra were obtained by preparing KBr samples and measured using a Perkin Elmer (Waltham, MA, USA) 1650 spectrometer. NMR-spectra were recorded with a Bruker DPX 400. The ^{1}H-NMR spectra were recorded at 400 MHz using $CDCl_3$ as a solvent, while the ^{13}C-NMR spectra were recorded at 100 MHz using the same solvent. XRD investigations were performed using a Siemens (Bruker (Billerica, MA, USA) AXS) D 5005 instrument with an OED Detector. The X-ray source was copper, and the frequency used was 0.154056 nm (Cu-K_α). Contact angles were measured using a Krüss DSA 10 with Drop Shape Analysis Software DSA 1 v 1.9. The testing liquids were distilled water and diiodomethane (Fluka (Geel, Belgium) AN 66880; purum > 98.0%), the volume of the drops was 10 µL and the deposition of a drop was carried out with 50 µL/min. Particle size was evaluated via dynamic light scattering (DLS) using a Zetasizer Nano S (Malvern Instruments Limited, Malvern, UK).

SEM micrographs were prepared with a Zeiss Supra 40 VP. Fluorine content of the products was determined by combustion of the dried samples according to the Schöniger method [39], followed by ion-sensitive fluoride determination.

To determine the surface roughness, atomic force microscopy (AFM) measurements were performed in contact imaging mode using an Agilent Technologies (Santa Clara, CA, USA) model 5500 beam-deflection atomic force microscope. Nanosensors with standard contact-mode cantilevers and a resonance frequency of 13 kHz, force constant of 0.2 N/m, length of 450 µm, width of 50 µm and thickness of 2 µm were used. The sampling resolution of all the images was set at 1024 data points with a scanning speed of 0.5 lines/s. All the captured images were processed using the open-source Gwyddion 2.49 software, and the arithmetic average roughness (Ra) was obtained by averaging the row/column roughness statistics of the whole image.

3.3. Plasma Etching

Plasma etching was carried out in a plasma chamber purged with argon/oxygen (1:1; v/v, flow rate: 25 ccm/min, pressure 10^{-4} bar). The RF-plasma operated at an excitation frequency of 13.56 MHz with a power of 40 W. Samples were treated for 10 min.

3.4. Preparation of Stöber Particles

Stöber particles were prepared by dissolving TEOS in ethanol, followed by the addition of specific amounts of ammonia and distilled water at room temperature. After initiating the hydrolysis, the sols were stirred for at least 16 h in a closed vessel. After preparation, the sols were concentrated by partially evaporating the solvent. For the preparation of 16 nm particles, 35.5 g (0.170 mol) TEOS was dissolved in 770 g EtOH, and 35.0 g of 25% aqueous NH_3 was slowly added.

3.5. Preparation of the Hydrophobic Silane

For preparation of the fluorinated silane (see Figure 1A), α,α,ω-trihydroperfluoroundecanol and 3-Isocyanatopropyltriethoxysilane were mixed in an equimolar ratio. Additionally, triethanolamine and tin octoate were added. The reagents were melted and kept at 80 °C in a nitrogen atmosphere for at least one hour. The product of the reaction was a colorless greasy wax, which was used without further purification. The fluorine content was measured as 44.1%, which is in good agreement with the theoretical value of 44.9%. An infrared spectrum of the product did not show any signals indicating residual isocyanato functions. The further NMR (Figure 6) and IR data are summarized below.

^{1}H-NMR: d_H (400 MHz; $CDCl_3$; Me_4Si): 6.06 (3t, 1H, $\underline{H}$-CF_2-, $^2J_{H\text{-}\underline{CF2}\text{-}CF2\text{-}}$ = 51.8; $^3J_{HCF2\text{-}\underline{CF2}\text{-}}$ = 5.04); 5.41 (b, 1H, -NH); 4.58 (m, 2H, -CF_2-$\underline{CH_2}$-O-); 3.83 (m, 2H, -O-$\underline{CH_2}$-CH_3); 3.22 (m, 2H, -NH-$\underline{CH_2}$-); 1.67 (m, 2H, -$\underline{CH_2}$-CH_2-CH_2-); 1.23 (t, 2H, -$\underline{CH_3}$); 0.64 (m, 2H, -$\underline{CH_2}$-Si).

^{13}C-NMR: d_C (100 MHz, de-coupled, $CDCl_3$, Me_4Si): 6.05 (-$\underline{C}H_2$-Si-); 16.44 (-$\underline{C}H_3$); 21.41 (-NH-$\underline{C}H_2$-CH2-); 42.09 ((-NH-$\underline{C}H_2$-CH_2-); 56.91 (Si-O-$\underline{C}H_2$); 58.39 (-CF_2-$\underline{C}H_2$-O-) $^2J_{FC\text{-}CH2\text{-}}$ = 104); 103–116 not resolved (H$\underline{C}F_2$-$\underline{C}_{10}F_{20}$-); 153.09 ($\underline{C}$=O).

IR (KBr molding film): ν_{max}/cm^{-1} 3348w (NH); 2978w (CH_3), 2932w (-CH_2-); 2889w (-CH_2-); 1732w (-C=O); 1533w (-C-N); 1445w (-CH_2-); 1380w (-CH_3); 1206s (-C=O); 1145s (-CF_2-); 1103s (-CF_2-); 770w (-CF).

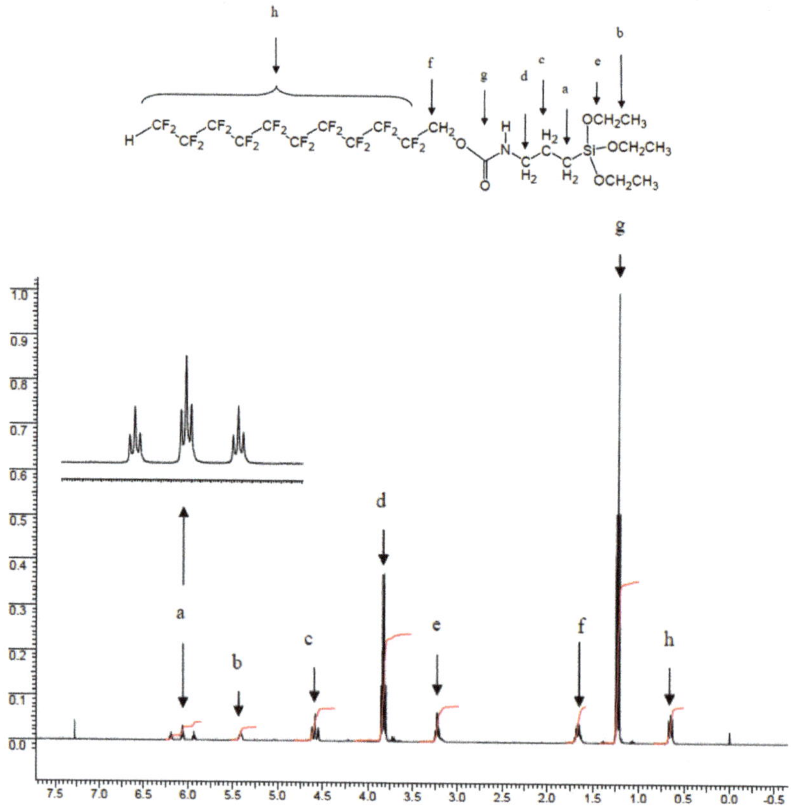

Figure 6. *Cont.*

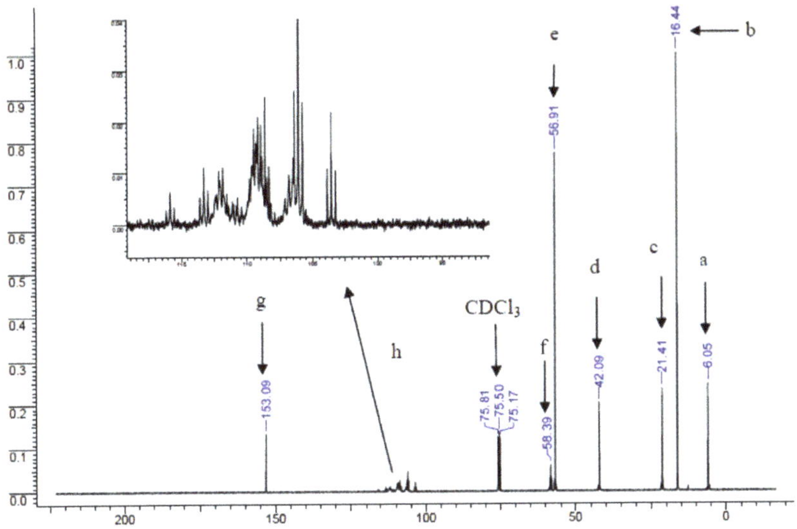

Figure 6. ^{1}H- and ^{13}C-NMR spectra of the fluorinated silane.

3.6. *Surface Grafting of Stöber Particles with the Hydrophobic Silane*

The surface-grafting of the silica particles was carried out by adding the previously synthesized alkoxysilane to the different silica sols (Scheme 1). The ratio of available silanol groups to added silane was approximately one-to-one. The silanol groups were determined via a method described by Gorlov et al. [37], which involves the chlorination of the silanol groups with thionyl chloride and a subsequent determination of the chloride content. The analysis of the fluorine content of the modified nanoparticles, after the removal of all solvents, showed a total fluorine content of 11.53% for the modified particles with a diameter of 15 nm. This value represents about 94.5% of the fluorine content calculated from the number of silanol groups theoretically available for condensation with the silane. Since the grafting with the comparably voluminous silane molecule will not remove all residual hydroxyl groups on the particle surface, and simultaneously not all silanol groups of the employed alkoxysilanes will crosslink due to steric reasons, an end-capping of the silanol functions can be carried out, e.g., to prevent an unwanted aging of the sols. The end-capping was performed using hexamethylsilazane.

3.7. *Preparation of a Wax Based on Luwax S and the Perfluoroethylene Alcohol (Fluowet EA 812 AC)*

To begin, 326.48 g (800 mmol) Luwax S (refined mixture of carbon acids from montan wax), 429.24 g (840 mmol) Fluowet EA 812 AC, 1.01 g sodium hypophosphite monohydrate and 3.78 g tetrabutyl orthotitanate were heated at 180 °C for 4 h in a three-necked flask equipped with a Dean–Stark apparatus. After 4 h, an additional 2 g tetrabutyl orthotitanate was added. The mixture was allowed to cool to 160 °C before adding 2 g NaCl solution. The dark mixture lightened immediately, yielding 703 g (97% of the theoretical yield) of a light-yellow wax with an acid number of 9.4 mg KOH/g.

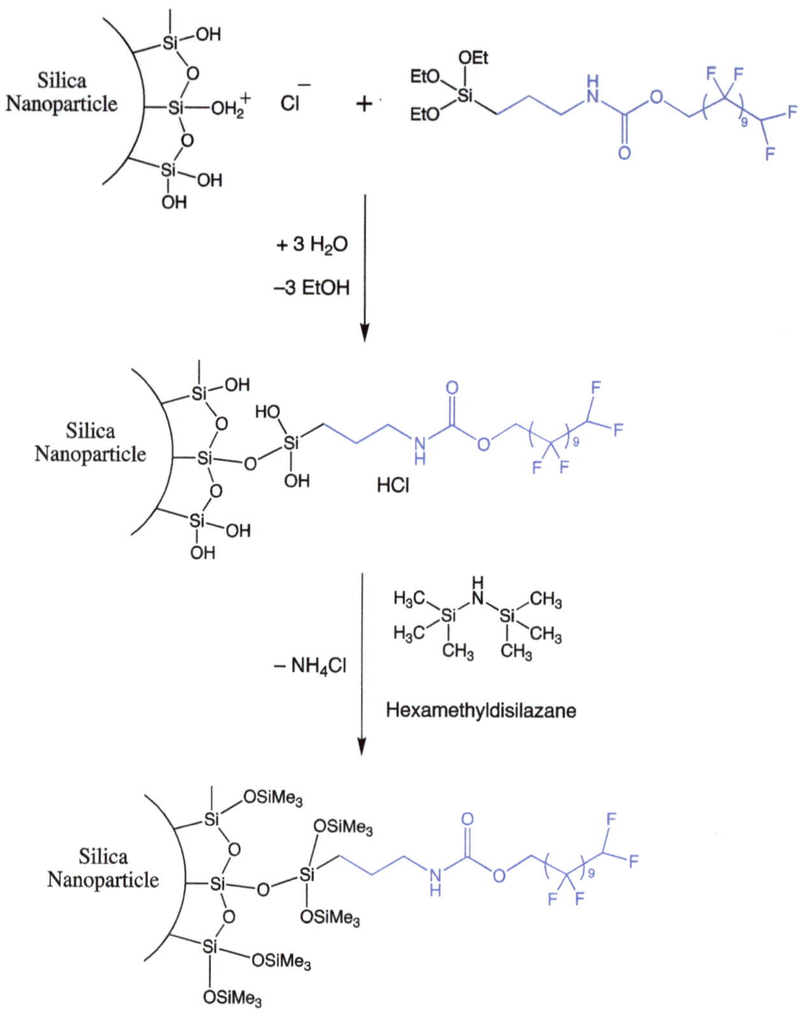

Scheme 1. Surface modification of silica nanoparticles with the synthesized fluorinated silane.

3.8. Preparation of an Alternative Wax Based on 2-Butyloctanoic Acid and the Perfluoroethylene Alcohol

A mixture of 3.64 g (10 mmol) 3,3,4,4,5,5,6,6,7,7,8,8,8-Tridecafluorooctanol and 2.00 g (10 mmol) of 2-Butyloctanacid were dissolved in 20 mL n-hexane. Then, 80 mg 4-dimethylaminpyridine and 2.84 g (13 mmol) di-tert-butyl dicarbonate were added to the mixture. The reaction mixture was stirred at 50 °C for three hours. Following this, 30 mL n-hexane was added before the mixture was transferred into a separating funnel and washed twice with 15 mL aqueous HCl (7 wt.%), twice with 15 mL aqueous $NaCO_3$ solution (10 wt.%) and twice with water. The organic phase was dried with 3 g of magnesium sulfate and then reduced using a rotary evaporator, before it was dried at ambient temperature and a pressure of 10^{-5} bar until no change in weight was observed. This process yielded 4.98 g (91% of the theoretical yield) of a colorless oil. Scheme 2 show the reaction.

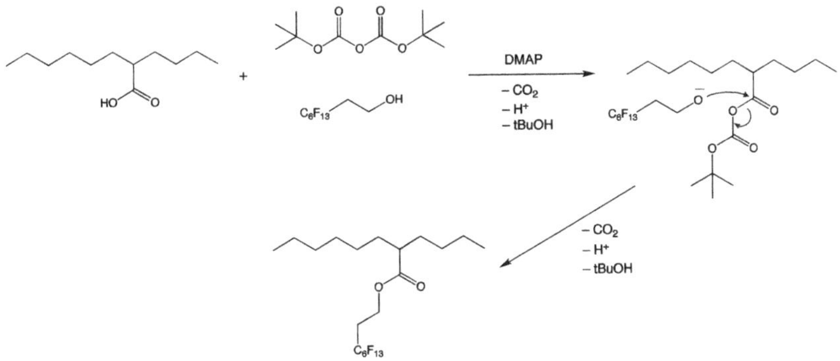

Scheme 2. Synthesis of the perfluoroethyl wax.

^{1}H-NMR (400 MHz; $CDCl_3$; ppm): 4.37 (m, 2H, C_6F_{13}-CH_2-$\underline{CH_2}$-; ^3J = 8); 2.51 (m, 2H, C_6F_{13}-$\underline{CH_2}$-CH_2-) 2.35 (2t, 1H, -$\underline{CH}$-COOH); 1.49 (m, 2H, C_3H_7-$\underline{CH_2}$-); 1.45 (m, 2H, C_5H_{11}-$\underline{CH_2}$-); 1.29 (m, 12H, -$\underline{CH_2}$-); 0.88 (m, 6H, -$\underline{CH_3}$).

^{13}C-NMR (100 MHz, decoupled, $CDCl_3$; ppm): 12.49 ($\underline{C}H_3$-C_3H_6-); 12.65 ($\underline{C}H_3$-C_5H_{10}-); 21.21 (CH_3-$\underline{C}H_2$-C_4H_8-); 21.24 (CH_3-$\underline{C}H_2$-C_2H_4-); 26.03 (C_4H_9-$\underline{C}H_2$-CH_2-); 27.85 (C_5H_{11}-$\underline{C}H_2$-CH-); 29.30 (C_2H_5-$\underline{C}H_2$-C_2H_4-); 28.25 (C_3H_7-$\underline{C}H_2$-CH-); 30.55 (C_3H_7-$\underline{C}H_2$-C_2H_4-); 31.03 (C_2H_5-$\underline{C}H_2$-C_3H_6-); 44.28 (CH Methine); 57.34 (Rf-CH_2-$\underline{C}H_2$-); 106–118 ($\underline{C}_6F_{13}$-); 174.87 (-$\underline{C}$=O).

IR (KBr liquid film; cm^{-1}): 2959 (s) ν_{as} -CH_3; 2932 (s) ν_{as} -CH_2-; 2861 (s) ν_s -CH_2-; 1740 (s) ν -C=O carboxylic acid esters; 1464 (m) ν_{as} -CH_3; 1458; ν_s -CH_2-; 1240 (s) ν -C-O- ester; ν -CF_3; 1206 (s) ν -CF_3; 1164 (w) ν -CF_2-; 1145 (m) ν -CF_2-; 733 (w) ν -CH_2- rocking; 709 + 700 (w) ν -CF.

3.9. Preparation and Investigation of the Composites

The composites were prepared by mixing the modified nanoparticles (directly from the sol) and the wax in the desired ratio. The mixture was stirred and heated to 80 °C until a clear solution was obtained. For application, the composite can be further diluted with the solvent (additional dispersant). For investigation of the resulting products, the mixture was simply dried to remove the solvent.

3.10. XRD Measurements

XRD measurements of the pure wax system are depicted in Figure 2 (blue), proving that the wax is a highly ordered system. In the investigated area, three signals can be observed between 2θ = 15 and 25. Notably, Lorenz investigated comparable systems [40] and described three identical peaks in this region (he investigated perfluorinated dendrimers based on C8-alkyl groups). Lorenz reported that these peaks indicate a layered structure of the perfluoro groups. The comparatively broad peak at 2θ = 17.8° can be explained by the interference of the perfluoroalkyl chains. Lorenz calculated a distance of 4.9 Å between the chains of the dendrimer he synthesized and the signal position identical to that in this work. However, Lorenz reported a half-width of 10 on the 2θ scale, whereas in this work, it was only 1. This indicates the formation of larger crystallites in this work (crystallite size is calculated to be 83.5 Å). The peaks at 21.4° and 24.0° are assigned to alkyl groups.

Author Contributions: Conceptualization, T.T., T.M.-G. and W.A.; methodology, U.H. formal analysis, T.T. and U.H.; investigation, U.H.; resources, E.S. and J.S.G.; data curation, U.H.; writing—original draft preparation, T.T.; writing—review and editing, T.T., T.M.-G. and U.H.; visualization, W.A.; supervision, E.S. and J.S.G. All authors have read and agreed to the published version of the manuscript.

Funding: This research received no external funding.

Institutional Review Board Statement: Not applicable.

Informed Consent Statement: Not applicable.

Data Availability Statement: The data are not publicly available.

Acknowledgments: This work is part of Uwe Holzdörfer's thesis, in which he cited an unknown author: *Experience is what is left when we became too old to experiment*. Unfortunately, Uwe passed away too early at the age of only 45. He was an awe-inspiring chemist, a kind soul and a dear friend.

Conflicts of Interest: The authors declare no conflicts of interest.

References

1. Wenzel, R.N. The Evaluation of Textile Waterproofing Agents. *Am. Dyest. Rep.* **1936**, *25*, 505.
2. Wenzel, R.N. The Evaluation of Textile Waterproofing Agents. III. Measurement of Waterproofness. *Am. Dyest. Rep.* **1936**, *25*, 598–601.
3. Wenzel, R.N. Resistance of solid surfaces to wetting by water. *Ind. Eng. Chem.* **1936**, *28*, 988–994. [CrossRef]
4. Peters, A.M.; Pirat, C.; Sbragaglia, M.; Borkent, B.M.; Wessling, M.; Lohse, D.; Lammertink, R.G.H. Cassie-Baxter to Wenzel State Wetting Transition: Scaling of the Front Velocity. *Eur. Phys. J. E* **2009**, *29*, 391–397. [CrossRef] [PubMed]
5. Cassie, A.B.D.; Baxter, S. Wettability of Porous Surfaces. *Trans. Faraday Soc.* **1944**, *40*, 546. [CrossRef]
6. Hipp, B.; Kunert, I.; Durr, M. Systematic Control of Hydrophobic and Superhydrophobic Properties Using Double-Rough Structures Based on Mixtures of Metal Oxide Nanoparticles. *Langmuir* **2010**, *26*, 6557–6560. [CrossRef] [PubMed]
7. Bhushan, B. Biomimetics Inspired Surfaces for Drag Reduction and Oleophobicity/Philicity. *Beilstein J. Nanotechnol.* **2011**, *2*, 66–84. [CrossRef]
8. Guo, Z.; Liu, W.; Su, B.-L. Superhydrophobic Surfaces: From Natural to Biomimetic to Functional. *J. Colloid Interface Sci.* **2011**, *353*, 335–355. [CrossRef]
9. Pilotek, S.; Schmidt, H.K. Wettability of Microstructured Hydrophobic Sol-Gel Coatings. *J. Sol-Gel Sci. Technol.* **2003**, *26*, 789–792. [CrossRef]
10. Rodriguez-Hernandez, J.; Muñoz-Bonilla, A.; Bousquet, A.; Ibarboure, E.; Papon, E. Environmentally Responsive Particles: From Superhydrophobic Particle Films to Water-Dispersible Microspheres. *Langmuir* **2010**, *26*, 18617–18620. [CrossRef]
11. Li, S.; Zhang, S.; Wang, X. Fabrication of Superhydrophobic Cellulose-Based Materials through a Solution-Immersion Process. *Langmuir* **2008**, *24*, 5585–5590. [CrossRef] [PubMed]
12. Zhang, G.; Wang, D.; Gu, Z.-Z.; Möhwald, H. Fabrication of Superhydrophobic Surfaces from Binary Colloidal Assembly. *Langmuir* **2005**, *21*, 9143–9148. [CrossRef]
13. Wagner, P.; Furstner, R.; Barthlott, W.; Neinhuis, C. Quantitative Assessment to the Structural Basis of Water Repellency in Natural and Technical Surfaces. *J. Exp. Bot.* **2003**, *54*, 1295–1303. [CrossRef] [PubMed]
14. Wang, L.; Shu, L.; Hu, Q.; Jiang, X.; Yang, H.; Wang, H.; Rao, L. Mechanism of Self-Recovery of Hydrophobicity after Surface Damage of Lotus Leaf. *Plant Methods* **2024**, *20*, 47. [CrossRef] [PubMed]
15. Li, C.; Wang, Y. *Wear, Fatigue, Creep, and Stress-Relaxation Behavior of Polymer Nanocomposites*; CRC Press: Boca Raton, FL, USA, 2010; p. 357.
16. Burris, D.L.; Boesl, B.; Bourne, G.R.; Sawyer, W.G. Polymeric Nanocomposites for Tribological Applications. *Macromol. Mater. Eng.* **2007**, *292*, 387–402. [CrossRef]
17. Rong, M.Z.; Zhang, M.Q.; Ruan, W.H. Surface Modification of Nanoscale Fillers for Improving Properties of Polymer Nanocomposites: A Review. *Mater. Sci. Technol.* **2006**, *22*, 787–796. [CrossRef]
18. Zimehl, R.; Textor, T.; Bahners, T.; Schollmeyer, E. From Colloids to Nanotechnology. *Progr. Colloid Polym. Sci.* **2004**, *125*, 49–53. [CrossRef]
19. Tao, J.; Liu, Y.; Li, M.; Li, Z.; Zhang, Y.; Song, X.; Yang, Q.; Guan, F.; Guo, J. Robust Superhydrophobic Composite Fabric with Self-Healing and Chemical Durability. *Small* **2024**, e2304894. [CrossRef]
20. Zhang, X.; Yang, J.; Zhang, W.; Ning, H.; Wang, H.; Liu, Z. Facile Preparation of Durable Superhydrophobic Coating with Microstructure Self-Repairing Function by Spraying Method. *Prog. Org. Coat.* **2024**, *187*, 108166. [CrossRef]
21. Zhou, S.; Jia, X.; Li, S.; Li, Y.; Tian, R.; Song, H. Epoxy/Polytetrafluoro-Wax Composite Coatings Demonstrate Remarkable Wear Resistance, Offering Rapid and Durable Cyclic Self-Healing Capabilities Facilitated by the "Sweating" Behavior of Polytetrafluoro-Wax. *Prog. Org. Coat.* **2024**, *188*, 108227. [CrossRef]
22. Celik, N.; Sahin, F.; Ozel, S.S.; Sezer, G.; Gunaltay, N.; Ruzi, M.; Onses, M.S. Self-Healing of Biocompatible Superhydrophobic Coatings: The Interplay of the Size and Loading of Particles. *Langmuir* **2023**, *39*, 3194–3203. [CrossRef] [PubMed]
23. Wei, J.; Yang, S.; Xiao, X.; Wang, J. Hydrophobic Solid Photothermal Slippery Surfaces with Rapid Self-Repairing, Dual Anti-Icing/Deicing, and Excellent Stability Based on Paraffin and Etching. *Langmuir* **2024**, *40*, 7747–7759. [CrossRef] [PubMed]
24. Wei, C.; Zhang, G.; Zhang, Q.; Zhan, X.; Chen, F. Silicone Oil-Infused Slippery Surfaces Based on Sol–Gel Process-Induced Nanocomposite Coatings: A Facile Approach to Highly Stable Bioinspired Surface for Biofouling Resistance. *ACS Appl. Mater. Interfaces* **2016**, *8*, 34810–34819. [CrossRef] [PubMed]

25. Fan, H.; Guo, Z. Robust Multi-Functional Slippery Surface with Hollow ZnO Nanotube Structures. *New J. Chem.* **2020**, *44*, 15483–15491. [CrossRef]
26. Lorusso, E.; Ali, W.; Hildebrandt, M.; Mayer-Gall, T.; Gutmann, J.S. Hydrogel Functionalized Polyester Fabrics by UV-Induced Photopolymerization. *Polymers* **2019**, *11*, 1329. [CrossRef] [PubMed]
27. Zhou, S.; Jia, X.; Tian, R.; Song, H. Self-Lubricating, Self-Healing and Anti-Corrosion Coating with Ultra-Short Run-in Period. *Tribol. Int.* **2024**, *196*, 109706. [CrossRef]
28. Trinh, B.; Owen, P.; vanderHeide, A.; Gupta, A.; Mekonnen, T.H. Recyclable and Self-Healing Natural Rubber Vitrimers from Anhydride-Epoxy Exchangeable Covalent Bonds. *ACS Appl. Polym. Mater.* **2023**, *5*, 8890–8906. [CrossRef]
29. Liu, J.; Li, J.-J.; Luo, Z.-H.; Zhou, Y.-N. Fast Room-temperature Self-healing Vitrimers Enabled by Accelerated Associative Exchange Kinetics. *Chem. Eng. J.* **2023**, *452*, 139452. [CrossRef]
30. Jiang, L.; Tian, Y.; Wang, X.; Zhang, J.; Cheng, J.; Gao, F. A Fully Bio-Based Schiff Base Vitrimer with Self-Healing Ability at Room Temperature. *Polym. Chem.* **2023**, *14*, 862–871. [CrossRef]
31. Matthies, L. Natural Montan Wax and Its Raffinates. *Eur. J. Lipid Sci. Technol.* **2001**, *103*, 239–248. [CrossRef]
32. Presting, W.; Kreuter, T. Zur Kenntnis Der Säuren Des Rohen Montanwachses Und Seiner Raffinate. *Fette Seifen Anstrichm.* **1962**, *64*, 695–701. [CrossRef]
33. Gooßen, L.J.; Döhring, A. A Convenient Protocol for the Esterification of Carboxylic Acids with Alcohols in the Presence of Di-t-Butyl Dicarbonate. *Synlett* **2004**, *35*, 0263–0266. [CrossRef]
34. Hassner, A.; Alexanian, V. Direct Room Temperature Esterification of Carboxylic Acids. *Tetrahedron Lett.* **1978**, *19*, 4475–4478. [CrossRef]
35. Seebach, D.; Hungerbühler, E.; Naef, R.; Schnurrenberger, P.; Weidmann, B.; Züger, M. Titanate-Mediated Transesterifications with Functionalized Substrates. *Synthesis* **1982**, *1982*, 138–141. [CrossRef]
36. Stöber, W.; Fink, A.; Bohn, E. Controlled Growth of Monodisperse Silica Spheres in the Micron Size Range. *J. Colloid Interface Sci.* **1968**, *26*, 62–69. [CrossRef]
37. Gorlov, Y.I.; Nesterenko, A.M.; Chuiko, A.A. The Chemisorption of Acid Chlorides on the Silica Surface. *Colloids Surf. A Physicochem. Eng. Asp.* **1996**, *106*, 83–88. [CrossRef]
38. Owens, D.K.; Wendt, R.C. Estimation of the surface free energy of polymers. *J. Appl. Polym. Sci.* **1969**, *13*, 1741–1747. [CrossRef]
39. Schöniger, W. Eine Mikroanalytische Schnellbestimmung von Halogenen und Schwefel in Organischen Substanzen. *Mikrochimica Acta* **1956**, 869–876. Available online: https://amuseum.de/chemie/Schoeniger/Publikationen/1956SchnellbestVonHalogenenUndSchwefelinOrgVerb.pdf (accessed on 5 August 2024). [CrossRef]
40. Lorenz, K.; Frey, H.; Stühn, B.; Mülhaupt, R. Carbosilane Dendrimers with Perfluoroalkyl End Groups. Core−Shell Macromolecules with Generation-Dependent Order. *Macromolecules* **1997**, *30*, 6860–6868. [CrossRef]

Disclaimer/Publisher's Note: The statements, opinions and data contained in all publications are solely those of the individual author(s) and contributor(s) and not of MDPI and/or the editor(s). MDPI and/or the editor(s) disclaim responsibility for any injury to people or property resulting from any ideas, methods, instructions or products referred to in the content.

molecules

MDPI

Review

Silane-Coupled Silica Nanoparticles Encapsulating Emitting Quantum Dots: Advancing Robust Phosphors for Displays and Beyond

Norio Murase [1,2,*] and Chunliang Li [3]

[1] Kansai Collaboration Center, National Institute of Advanced Industrial Science and Technology (AIST), Ikeda 563-8577, Osaka, Japan
[2] Quantum Materials Technology Co., Ltd. (QMT), 2-22-11 Obana, Kawanishi 666-0015, Hyogo, Japan
[3] School of Materials Science and Engineering, Tianjin University of Technology, Xiqing, Tianjin 300384, China
* Correspondence: n-murase@qmt.co.jp or n-murase@aist.go.jp or nmurase@icloud.com

Abstract

Colloidal quantum dots (QDs) are semiconductor crystals a few nanometers in size. Due to their vibrant colors and unique photoluminescence (PL), QDs are widely utilized in displays, where barrier films provide essential shielding. However, one of the primary challenges of QD applications remains achieving sufficient robustness while keeping costs low. Over the past two decades, significant progress has been made in the encapsulation of QDs within silica matrices, aiming to preserve their original PL properties. Research efforts have evolved from bulk forms to thin films. Silica nanoparticles containing multiple embedded QDs have emerged as particularly promising candidates for practical applications. This review highlights recent advancements in silica-based QD encapsulation, incorporating findings from both the authors' investigations and those of other research groups within the field. Silica glass possesses inherent shielding capabilities, but silane coupling agents such as (3-aminopropyl)trimethoxysilane and (3-mercaptopropyl)trimethoxysilane tend to negatively impact this functionality when they are used alone, partly because of the limited formation of a well-developed glass network structure. However, when judiciously controlled, they can serve as mediators between the QD surface and the surrounding pure silica glass matrix, helping to preserve PL properties and control the morphology of silica particles. This review discusses the potential for achieving exceptional shielding properties through sol–gel glass fabrication at low temperatures, utilizing both tetraethoxysilane and other silane coupling agents.

Keywords: sol-gel; silane-coupling agent; colloidal quantum dot; photoluminescence; protection

1. Introduction

Silica, particularly in its amorphous form (silica glass), plays a crucial role in the semiconductor industry because of its exceptional protective, insulating properties and adaptability in shaping [1]. This distinctive nature may also be indispensable as a matrix for quantum dot (QDs) emissions. Throughout this article, amorphous silica or silica glass is collectively referred to as "silica". Since the QDs discussed here are synthesized in solution (colloidal QDs), the silica is similarly prepared in solution via the sol–gel method to ensure compatibility. In contrast to the vacuum-based processes commonly employed in the semiconductor industry, the solution-phase chemical reactions required to form protective layers on QDs are dependent on the use of silane coupling agents. This contribution to the Special Issue on hybrid organic/inorganic (O/I) sol–gel-derived nanocomposite systems summarizes accumulated knowledge on the preparation and evaluation of sol–gel-derived silica-encapsulated QD emissions. Relevant studies from other research groups are also critically examined to provide a comprehensive overview.

It has been more than a decade since the first commercial release of televisions utilizing QDs as photoluminescence (PL) sources, commonly referred to as QD TVs [2]. The primary application of QDs continues to be as phosphors in display technology, with the overall market size reportedly valued at billions of dollars annually [3]. Another potential use is in biomedical tagging, where QDs function as fluorescent reagents [4–6]. However, both applications still demand enhanced robustness in QD formulations. Following this introductory section, Section 2 presents a concise overview of the related research areas, with emphasis on colloidal QDs and their protective materials, and a brief outline of the structure and scope of this article. This discussion aims to facilitate a better understanding of subsequent arguments. Section 3 outlines the basic principles of the preparation of silica matrices with encapsulated emitting QDs, along with the results observed in bulk and thin-film forms. Given their promising potential for applications, silica nanoparticles with encapsulated QDs are particularly noteworthy. Accordingly, the subsequent Section 4 explores four distinct nanoparticle preparation methods, categorized based on two types of QDs (hydrophilic and hydrophobic), and two sol–gel approaches (reverse micelle and the Stöber method). Related research from other groups is also provided. Among these, the preparation of silica nanoparticles with dozens of encapsulated hydrophobic QDs using the Stöber method yielded the most promising results, with the aid of a silane coupling agent. As this particular method demonstrated significant advantages, Section 5 delves deeper into preparations and evaluations of these silica particles, despite their classification under hydrophobic QD encapsulation by the Stöber method in Section 4. Recent advancements, applications, and challenges in the field are discussed in Section 6. Finally, Section 7 offers concluding remarks with perspectives on future developments. Table 2, discussed later, is intended to facilitate understanding of the positioning and classification outlined in this review.

2. Background of Related Fields and Article Outline

2.1. Colloidal Quantum Dots

Colloidal QDs are semiconductor nanocrystals synthesized in solution, typically measuring a few nanometers in diameter. Their size places them between atomic clusters and bulk materials, as illustrated in Figure 1, with each QD crystal consisting of approximately 10^2–10^4 atoms per particle. The quantum size, also known as quantum confinement, causes the band gap to widen as the crystal size decreases, actually originating from the uncertainty principle, which is schematically depicted in Figure 2 [7]. Emitting QDs generally consist of three components: a core, a shell, and surface ligands (surfactants), as shown in Figure 3. To achieve narrow and intense PL, the band gap of the shell must be larger than that of the core. The ligand structure also influences the solubility properties of QDs. Hydrophilic QDs are typically functionalized with ligands containing short carbon chains (fewer than three–four carbon atoms). Conversely, hydrophobic QDs are stabilized during synthesis using ligands with longer carbon chains (generally exceeding seven carbon atoms). Both hydrophilic and hydrophobic colloidal QDs are precipitated upon the addition of lower alcohols such as ethanol and methanol. This purification method is generally used during synthesis. Following this process, ligand exchange is frequently performed post-synthesis to tailor surface properties for specific downstream applications. Given their small size, a large proportion of QD atoms are surface-exposed, making surface chemistry a critical factor in determining PL efficiency.

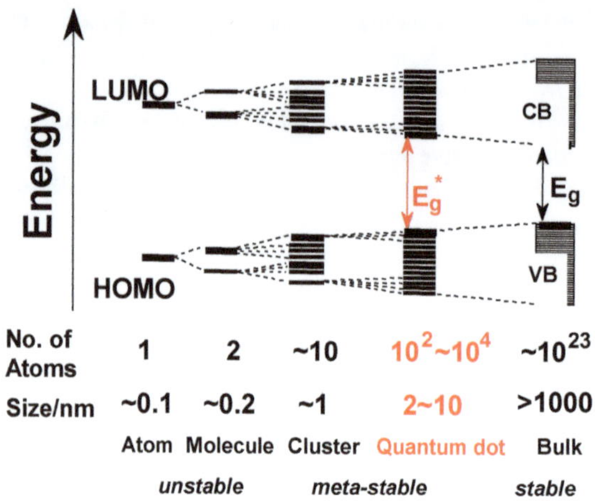

Figure 1. The relationship between particle size and band gap in semiconductors. As particle size decreases, quantum confinement effects become more pronounced, leading to an increase in band gap energy. This phenomenon plays a crucial role in tuning optical and electronic properties in nanoscale semiconductor materials, such as QDs.

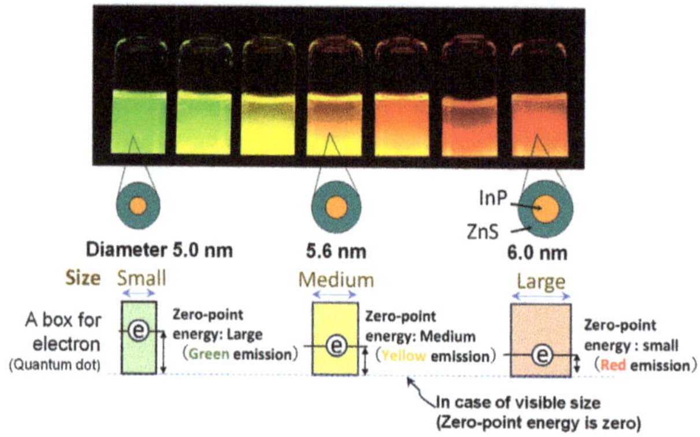

Figure 2. Schematic representation of the quantum size effect. As the QD size decreases, quantum confinement effects become more pronounced, leading to a widening of the bandgap, according to the value of zero-point energy originating from the uncertainty principle [7] (Courtesy of AIST, Japan). This effect was first observed and precisely explained by two Nobel laureates in chemistry: Dr. A. I. Ekimov [8], and Dr. L. E. Brus [9].

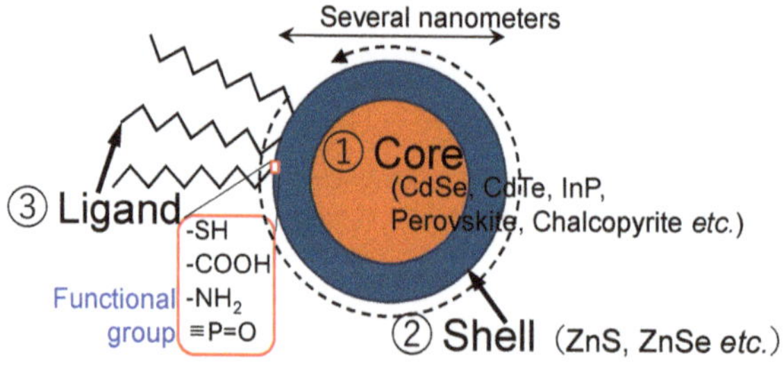

Figure 3. Schematic representation of the structure of emitting QDs, consisting of three primary components:①the core, which defines the fundamental optical and electronic properties;②the shell,

which enhances stability and emission efficiency; and ③ the surrounding ligand molecules, which facilitate dispersion and surface interaction. The functional group of PO in the figure is from trioctylphosphine oxide. In certain cases, the shell is embedded within the core structure during synthesis, as exemplified by aqueous CdTe QDs [10]. Perovskite QDs exhibit strong PL originating predominantly from their core structure; however, the presence of an outer shell significantly improves their structural robustness and stability [11].

The most typical hydrophilic emitting QD widely studied is CdTe, capped with low molecular weight thiols such as thioglycolic acid [12]. In the case of hydrophobic QDs, the most investigated and highest-quality QDs are those that are CdSe-based, pioneered by the success of hot injection synthesis [13]. Their shells, typically composed of CdS, ZnSe, or ZnS, enable a remarkably narrow PL spectral width (approximately 25 nm or less in full width at half maximum, FWHM), ensuring exceptional color fidelity in display applications. However, cadmium-based QDs are being phased out due to RoHS restrictions [14]. As an alternative, InP-based QDs currently stand out as the sole viable, environmentally friendly binary QDs capable of emitting visible light through quantum size effects [15]. Despite early awareness of their potential, research on InP QDs lagged behind that of CdSe by more than a decade, largely due to the challenging reaction conditions inherent to III–V semiconductor systems and the difficulty of forming an optimized interface between InP cores and II–VI shells (such as ZnSe or ZnS). After multiple trials [16–18], InP QDs finally achieved nearly 100% PL quantum yield (PLQY) [19,20]. Currently, high-end displays have begun adopting InP-based QDs as an alternative to CdSe-based QDs, despite their relatively broad PL spectral width (typically exceeding 35 nm FWHM).

In addition to traditional binary QDs (II–VI and III–V), two-types of ternary QDs, namely chalcopyrite (I-III-VI_2 or II-IV-V_2, such as $CuInS_2$, $AgInS_2$, and sometimes quaternary) and perovskite ($CH_3NH_3PbX_3$ or $CsPbX_3$, where X = Cl, Br, I), have emerged. These materials show narrow and intense PL (typically less than 25 nm), comparable to CdSe-based QDs.

Non-toxic chalcopyrite QDs were first reported as Zn-$CuInS_2$ in 2006 [21]. These QDs exhibited PL accompanied with strong defect emissions, which likely originated from deviations in stoichiometric composition due to the limited number of atoms within individual QDs. In the case of $AuInS_2$ QDs, the formation of a specific shell structure significantly reduced the defect emission [22]. However, the suppression was not entirely complete. Ongoing efforts continue to improve the PL properties and processability of these QDs [23]. Several studies have also reported the silica encapsulation of these QDs to enhance their stability, as briefly outlined in Section 2.4.

Lead-based perovskites represent a distinct class of emitting QDs. Although their synthesis procedures are broadly similar to those used for traditional hydrophobic II-VI, they are highly ionic. Unlike other QDs, their emission is largely insensitive to the surface conditions [24], although the shell formation is effective for structural protection [11]. Currently, stability remains a significant drawback for these QDs as well. Consequently, extensive research has been conducted on their encapsulation within silica matrices, as outlined in Section 2.3. Lead (Pb), a regulated heavy metal, presents environmental and safety concerns in consumer applications. However, ongoing advancements in solar cell technologies, employing the same perovskite compositions in bulk rather than QD form, have contributed to a relaxation of regulatory constraints for QD-based consumer products.

The remaining class consists of single-component (unary, group IV) QDs. Carbon-based QDs are quite sensitive to the surface, and their PL is typically characterized by a broad emission profile [25]. Similarly, silicon QDs generally show broad PL spectra, and both systems continue to face significant challenges in terms of stability [26]. To date, no substantial studies have been identified regarding the incorporation of these unary QDs into silica matrices.

2.2. Matrices for QD Encapsulation

As discussed above, QDs require coating and protection with another transparent material to preserve their distinctive PL properties. Table 1 summarizes research efforts dedicated to the shielding of emitting QDs, including contributions from key producers [27–30], the authors' own investigations, and various polymer-based approaches. Attempts to encapsulate QDs within crystalline materials have largely been unsuccessful. Since QDs themselves are crystals, achieving a stable interface with another crystalline matrix is challenging, particularly at high dispersion concentrations. This often leads to surface defects that reduce PLQY. Among transparent amorphous materials, polymer and glass (such as silica glass) stand out as viable encapsulation matrices. Due to the significantly lower oxygen permeability of glass compared to polymers [31], glass is widely regarded as the optimal choice for QD encapsulation. Consequently, thousands of publications have explored sol–gel-derived silica as a protective encapsulation medium. Representative examples will be discussed following categorical classification later in this work. However, due to the complexity of sol–gel processing and the relatively lower performance of the resulting materials, most studies in this area have not been actively pursued. Nonetheless, the continued and substantial contributions made by the authors' research group offer a valuable foundation for interpreting progress in this field. Accordingly, the authors encourage readers to consider these findings as a core framework of this review for understanding the development and future prospects of silica-encapsulated QD technologies, as elaborated in the following subsection.

Table 1. Properties of transparent matrices for encapsulation of emitting QDs [(a)].

Matrix \ Property			**Shielding Ability (Photostability)**	**PLQY**	**Main Producer**
Organic	Amorphous	Polymer	−	+	General
Inorganic		**Silica** [(b)]	+	+/−	The authors
	Crystal	Zeolite	+	−	Korean group [27]
		Alumina	+	−	Crystalplex Inc., USA [28], Related papers [29,30]

[(a)] A plus sign (+) in the table indicates a positive or promising outcome, while a minus sign (−) denotes a negative or unfavorable result. [(b)] An inorganic and amorphous matrix shown in blue is preferable to encapsulate the crystalline QDs with preventing the intrusion of oxygen. We therefore have selected silica as shown in a thin yellow background.

2.3. Outline of the Article

Table 2 summarizes the methods for the encapsulation of emitting colloidal QDs in silica matrices, which are categorized from letters A to F, corresponding to the respective sections discussed in this article.

Briefly, two types of hydrophilic and hydrophobic QDs, immediately after synthesis, were encapsulated using the sol–gel process of reverse micelle or the Stöber method, and other approaches. Given that current display technologies primarily employ binary QDs based on CdSe and InP (with ZnSe-based variants used in limited demonstration stages), research activities have largely concentrated on these materials. The results of authors' studies are compared with existing reports, and it is concluded that F, corresponding to the encapsulation of hydrophobic QDs into silica particles assisted by silane coupling agents, yields the most favorable outcomes.

2.4. Silica Encapsulation of Ternary QDs

A selection of representative studies on the silica encapsulation of ternary QDs is discussed in this subsection. The results are interpreted with reference to the encapsulation categories defined in Table 2.

Chalcopyrite QDs ($CuInS_2$-based) have been encapsulated within silica nanoparticles using tetraethoxysilane (TEOS) via reverse micelle methods. Both hydrophilic (E) [32] and hydrophobic (A) [33] variants have been investigated, though only the latter was reported to have a PLQY of 1.5%. Stability assessments indicated limited improvement upon encapsulation when compared to the pre-encapsulated state. Additionally, silane coupling agents, including (3-aminopropyl)trimethoxysilane (APS), have been employed for $CuInS_2$ QDs (hydrophobic version of No. B) [34–36]. The silica precursor used in reference [34] was not a conventional silane coupling agent, but it exhibited reaction behavior closely resembling such agents. Overall, PLQYs achieved using this approach are generally superior to those obtained via the reverse micelle method. However, the degree of glass network formation appears suboptimal, as indicated by the absence of noticeable shrinkage in the resulting monoliths upon completion of the preparation process. The stability of silica-encapsulated chalcopyrite QDs exceeds that of their pristine counterparts [35]. For $AgInS_2$-based QDs, only one instance of successful silica encapsulation has been identified, involving growth within a molten glass matrix. This approach yielded broad PL emission across the visible spectrum [37].

Perovskite QDs have been more actively integrated into silica matrices compared to other QD systems. The predominant approach involves standard sol–gel methods utilizing silane coupling agents such as APS [38–40] or alkoxides with four functional groups such as TEOS [41] or tetramethoxysilane (TMOS) [42,43], while the latter exhibits a significantly higher hydrolysis rate than that of TEOS. These encapsulation strategies are classified under B, targeting hydrophobic QDs. A maximum PLQY of 78% was reported for QDs encapsulated using APS [38]. However, stability assessments in these studies were limited to comparisons with pristine QDs or those embedded in polymethyl methacrylate (PMMA) [38]. In another work, hydrophobic QDs were encapsulated using a modified Stöber method (F), with stability evaluated relative to their pristine counterparts [44].

Commercial mesoporous silica matrices have been employed to incorporate colloidal QDs [45–47]. However, no comparative data on stability with other solid matrices has been reported. Among the available studies, the most promising one appears to be the synthesis of $CsPbBr_3$ QDs within a molten glass matrix, achieving a PLQY of 42% [48]. Although the size distribution of QDs in the glass is broad, these QDs are stable in air for over one month. A comprehensive investigation has described the growth mechanism of these QDs in glass matrices at elevated temperatures, such as 600 °C, over extended durations exceeding 10 h. The study combines experimental results with theoretical insights into particle growth mechanisms [49]. Nevertheless, a significant limitation of this method is the inability to purify and further modify the QDs post-incorporation, such as enhancing PLQY after synthesis [50]. Therefore, the authors have advocated for research into dispersing colloidal QDs into silica matrices through sol–gel approaches following colloidal synthesis.

Here, it is worth noting that most of the encapsulation methods for ternary QDs emerged in the 2010s. These early research attempts failed to result in systematic or quantitative studies, limiting their impact on methodical development and comparative analysis.

Table 2. Overview of binary QD emission incorporated into silica glass matrices using various sol–gel methods. The encapsulation approaches are categorized from A to F, corresponding to detailed discussions provided in relevant sections of this article. The non-English term (nano glass in Japanese) in the image of category No. B is the name of research project.

Category No.	Quantum Dot	Interaction with Water	Silane-Coupling Agent	Method	Morphology of Silica Glass	Comments	This Article	Main Year	References
A	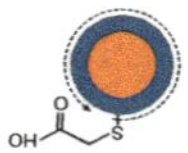 CdTe, ZnSe, & InP-based	Hydrophilic	—	Reverse micelle	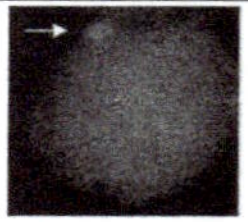 Nanoparticle	Not incorporated into glass spheres.	Section 4.1	2004	[51–53]
B			APS	(Normal sol-gel)	 Bulk	First successful incorporation into glass with high PLQY; however, the glass dissolved upon exposure to hot water.	Section 3.1	2004	[54–58]
C			APS	Layer-by-layer	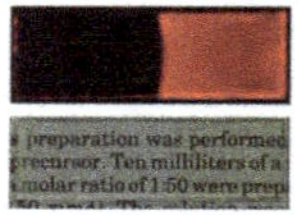 Thin film	High QD loading achieved; however, the resulting film was fragile.	Section 3.2	2005	[59]
D			MPS	Stöber	 Nanoparticle	Porosity was characterized; QDs exhibited a broad FWHM.	Section 4.2	2010, 2020	[60–63]
E	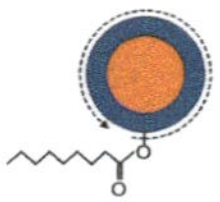 CdSe, & InP-based	Hydrophobic	—	Reverse micelle	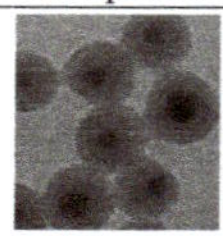 Nanoparticle	Single QD per silica particle achieved; however, dispersion concentration remained insufficient.	Section 4.3	2006	[64–69]
F			MPS	Modified Stöber	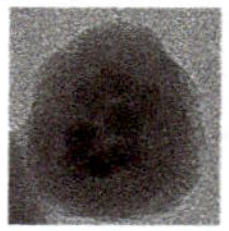 Nanoparticle	Multiple QDs per silica particle; extensively investigated due to several applications.	Section 4.4, and Sections 5 and 6	2010–2025	[70–77]

3. Incorporation of QDs in Bulk and Thin-Film Silica Derived from Silane Coupling Agents

3.1. Bulk Incorporation of QDs Using APS

Following the successful synthesis of hydrophobic CdSe-based QDs [78], these QDs were incorporated into polymer matrices molded within glass tubes for industrial applications [79]. However, the reported PLQY remained relatively low, ranging between 20 and 40%. Around the same time, researchers introduced a modified sol–gel method to disperse QDs within silica [54]. However, the prepared matrix is seemingly a gel, not a solid. Unlike polymer-based dispersions, the sol–gel approach requires sufficient reaction time to achieve uniform solidification, a critical factor in optimizing the method.

Building upon the fundamental properties of colloidal emitting QDs, a series of investigations was conducted to evaluate their stability and optimize incorporation into silica matrices. Generally, the radius (r) of QDs in solution grows or dissolves according to the equation [80]

$$\frac{\mathrm{d}r}{\mathrm{d}t} = \frac{K}{r}\left(\frac{1}{r_\mathrm{c}} - \frac{1}{r}\right), \quad (1)$$

where K is a constant determined by the solution temperature, the diffusion coefficients of the components, and the specific surface energy of QDs. When the radius r of QD is smaller than a critical value r_c, it dissolves and finally disappears into the solution. The dissolution process leads to a rapid and significant decrease in PLQY. The constituent atoms or clusters of such QDs are used to grow other QDs larger than r_c. This phenomenon is known as Ostwald ripening. To create bulk silica glasses incorporating QDs, aqueous CdTe-based QDs stabilized with thioglycolic acid (TGA) were initially selected. The core ligand structure of TGA features a single carbon skeleton, offering minimal steric hindrance and facilitating uniform dispersion within the sol–gel matrix.

Based on Equation (1), a tailored protocol was developed to preserve the initial PLQY of QD after the incorporation into silica matrices.

Selection of Silane Coupling Agent: The amino-functionalized silane (APS) was chosen to facilitate stable QD dispersion. This selection leveraged the carboxylic acid groups present on TGA-stabilized QDs. The amine functionality not only exhibits affinity toward carboxyl groups facilitating coordination with QD surface ligands, but also serve as a catalyst for gel formation during the sol–gel process [54].

Dispersion Strategy: Constituent molecules and ions including TGA and Cd^{2+} were dispersed in APS solution to minimize the dissolution of QDs during the sol–gel process. In other words, Equation (1) indicates that the dissolution of QDs is effectively suppressed when the particles are surrounded by a sufficient concentration of constituent molecules and ions. This protective environment helps stabilize the QDs by minimizing surface degradation and maintaining their PL properties.

Optimized Timing for QD Addition: To mitigate time-dependent dissolution, QDs were introduced into the sol phase immediately prior to gelation. This timing ensures rapid immobilization within the forming silica network, effectively preventing QD dissolution and preserving their structural integrity and PL properties.

As a result, the first bulk glass containing dispersed emitting QDs was successfully fabricated, maintaining an initial PLQY of ~40%, as illustrated in Figure 4. Due to the release of methanol during the hydrolysis of APS, which acts as a poor solvent for QDs, the achievable dispersion concentration of QDs within the resultant silica matrix remained relatively low, typically in the order of 10^{-5} mol/L [55]. As the brightness of QDs is directly proportional to the dispersion concentration in appropriate conditions, higher concentrations are desirable for practical applications.

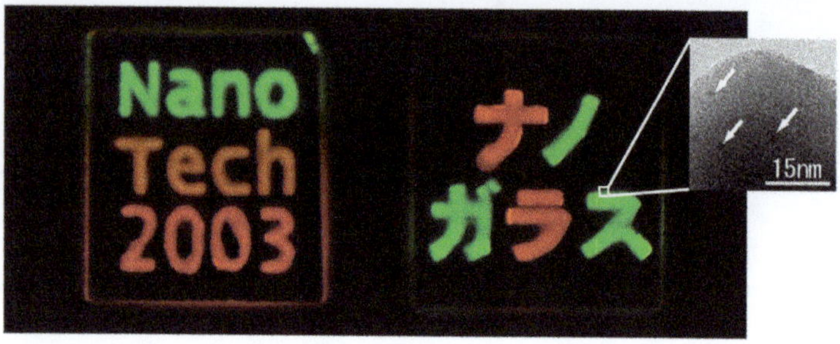

Figure 4. Fluorescent color images of the prepared bulk-type glass containing QDs under UV lamp excitation at 365 nm. The phosphor is selectively adhered to patterned grooves on the surface of the glass substrates, enhancing localized emission [55]. The non-English term (nano glass in Japanese) is the name of research project sponsored by a funding agency (NEDO). The color variation is due to the different size of aqueous CdTe-based QDs.

Similar results were reported more recently for hydrophobic CdSe QDs. Although TEOS was used in combination with other silane-coupling agents, its molar ratio was maintained below 10% [56]. Consequently, the resultant monoliths exhibit characteristics comparable to those of APS-based matrices, wherein the initial PLQY is relatively well preserved. Notably, the dispersion concentration of QDs achieved in this case is estimated to be approximately two orders of magnitude higher (5 wt%) than that reported in the earlier APS-based approach.

In the case of InP-based QDs, hydrophobic variants were first hydrophilized and subsequently incorporated into silica matrices synthesized from silicon compounds, exhibiting behavior similar to silane coupling agents. The PLQY was relatively well maintained, decreasing from an initial 52% to 27% upon encapsulation. However, insufficient development of the silica network is suggested by the absence of characteristic cracks which typically result from condensation and shrinkage in bulk materials. The stability was evaluated against PMMA [57]. In a subsequent effort, four-functional alkoxides (TMOS and TEOS) were employed for encapsulating the same hydrophobic QDs. In the case of TMOS, a high PLQY was reported, likely reflecting the properties of QDs embedded in a gel-like matrix [58].

3.2. Layer-by-Layer (LbL) Incorporation of QDs Using APS

To enhance QD incorporation into silica matrices, a layer-by-layer (LbL) deposition method was employed using APS [59], building upon the previously discussed bulk encapsulation results. In this approach, a glass slide was sequentially dipped into APS solution, TGA-containing solution, and QD solution, forming transparent QD layers between the glass layers, as shown in Figure 5a. The inset of Figure 5b demonstrates a linear increase in absorbance with each successive dipping cycle, indicating progressive QD accumulation. Using the molar extinction coefficient of QDs [81], the concentration was estimated to be 10^{-2} mol/L, approaching the threshold of concentration quenching. At such high concentrations, the PL peak wavelength is red shifted with a narrowing of spectral width due to the re-absorption of PL. In this phenomenon, PL emitted from smaller QDs is reabsorbed by adjacent larger QDs, leading to selective emission primarily from the larger QDs. Consequently, the observed PL spectrum shifts toward the red end, with a reduction in full width at half maximum (FWHM). These observations confirm that APS facilitates effective QD incorporation, making it a preferred choice for both the bulk glass incorporation of hydrophobic QDs [82] and thin-film applications [83].

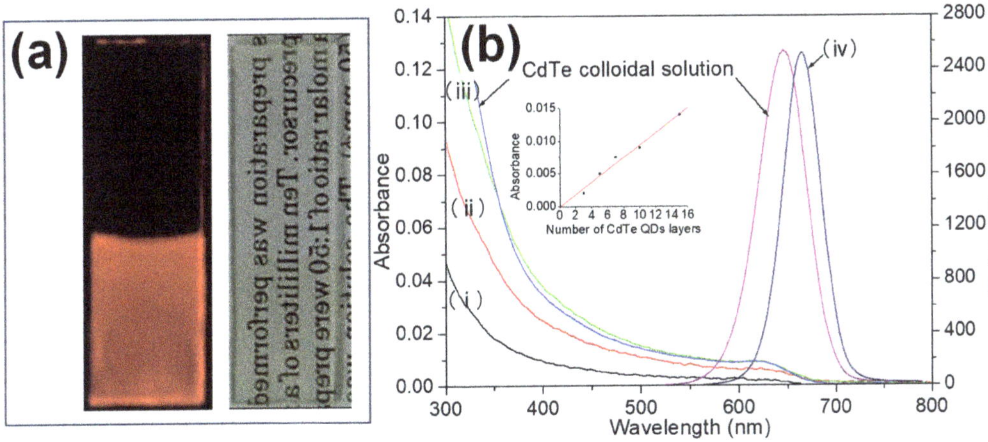

Figure 5. (**a**) PL image of the self-assembled films irradiated with 365 nm UV light prepared using APS and CdTe colloidal solution (left). A photograph of the sample taken under ambient room lighting (right). (**b**) PL and absorption spectra of the films. Spectra (i), (ii), and (iii) correspond to absorption measurements for samples with 3, 5, and 10 CdTe layers, respectively. Spectrum (iv) represents the PL response of the film with 10 CdTe layers. The inset in (**b**) illustrates the absorbance at the first absorption peak as a function of the number of dipping cycles. For comparison, the PL and absorption spectra of diluted QD colloidal solutions are also provided [59].

When the applied sol–gel process does not require stirring, such as the LbL process, the QDs do not suffer from dissolution according to the mechanism outlined in Equation (1). This allows for a significant increase in dispersion concentration while largely preserving the initial PLQY. However, glasses prepared using only three-functional alkoxides, such as APS, lack a robust glass network both for bulk and LbL. As a result, they exhibit insufficient structural integrity and eventually dissolve completely in hot water.

4. Incorporation of QDs into Sol–Gel-Derived Silica Nanoparticles

As research progressed, it became evident that silica nanoparticles with encapsulated QDs are particularly promising for both display and biological tagging applications. Among encapsulation methods, the step-by-step grafting of alkoxide molecules [84,85] was limited in effectiveness, since this method cannot be used to achieve thicknesses beyond 10 nm. However, excluding this approach, the sol–gel encapsulation method can generally be classified into two main categories: the reverse micelle method and the Stöber method. The reverse micelle method uses a water-in-oil emulsion where the alkoxide initially disperses in the continuous oil phase but gradually goes into the water droplet phase upon hydrolysis. In the Stöber method, the typical alkoxide TEOS reacts with an alkaline aqueous solution of abundant ethanol to form hydrolyzed TEOS, which later condenses onto colloidal QDs, resulting in the formation of nanosized silica particles with encapsulated QDs. As described above, the QDs are of two types: hydrophilic and hydrophobic. Therefore, the encapsulation methods result in four distinct types. The following sections detail each method separately.

4.1. Hydrophilic QDs Incorporated into Silica Nanoparticles via Reverse Micelle Method

For hydrophilic QDs, the reverse micelle method leverages water droplets within the water-in-oil microemulsion as their dispersion medium. Initially, hydrophobic TEOS molecules remain in the continuous oil phase but gradually migrate into the water droplet phase upon hydrolysis. Subsequently, hydrolyzed TEOS condenses in an alkaline environment, forming silica nanoparticles. The aqueous QDs (TGA-capped CdTe QDs) were initially expected to be incorporated into the silica particles by this protocol. However, the transmission electron microscope (TEM) image (Figure 6) revealed that the QDs were pushed away during the silica network formation, preventing effective encapsulation [51]. To overcome this challenge, a hydrophilic vitreous layer was introduced onto the QD surface prior to sol–gel processing, successfully enabling multi-QD incorporation into nanoscale silica particles. This refinement allowed red-emitting QDs to maintain a PLQY of 65% post-encapsulation [52].

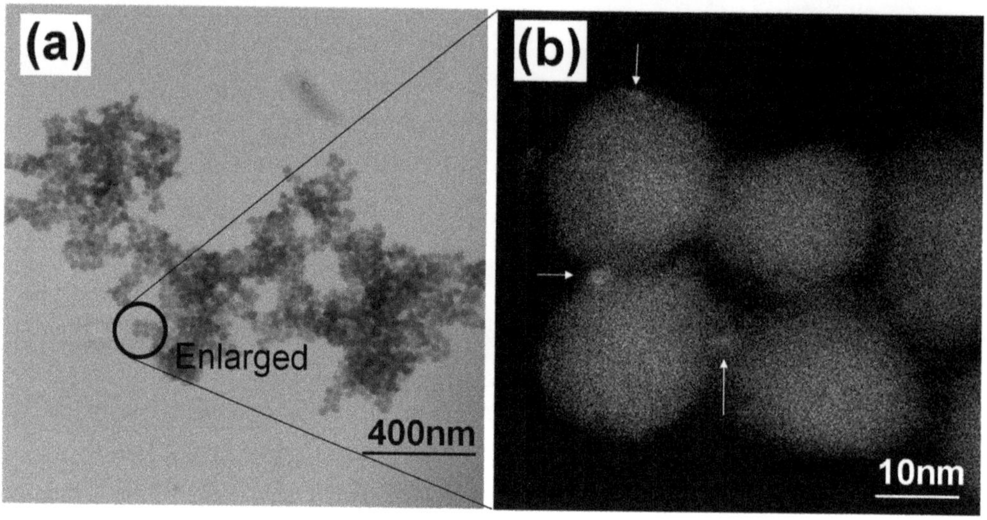

Figure 6. (**a**) Transmission electron microscope (TEM) image of silica spheres synthesized using a reverse micelle method with aqueous QDs in bright field mode. (**b**) High-angle annular dark field (HAADF) image of the region indicated by a circle in (**a**), highlighting locations where Cd is detected, as marked by the arrows [51].

In contrast, aqueous CdS QDs synthesized within reverse micelle droplets were successfully encapsulated using TEOS, without the need for additional surface modifications in advance [53]. This discrepancy appears to stem from CdS QDs being inherently stabilized in an alkaline environment (rich in hydroxyl ions) without requiring external ligands. However, PL properties for this system were not reported.

4.2. Hydrophilic QDs Incorporated into Silica Nanoparticles via the Stöber Method

In the early stages of research, hydrophilic QDs such as CdTe stabilized by short-chain thiols, or CdSe stabilized by sodium citrate, were incorporated into silica particles using sodium silicate. These silica particles containing multiple QDs were referred to as "Raisin Bun-type" silica composites. To further enhance their structure, the Stöber method was used to coat the silica spheres, increasing their size beyond 200 nm. However, at this stage, PLQY remained low, and the particles were primarily used to construct three-dimensional photonic crystals [60].

To improve encapsulation efficiency, a two-step process was developed for integrating multiple TGA-capped CdTe QDs into silica particles [61].

Step 1—QD Assembly Using MPS: Since thiol ligands exhibit the strongest binding energy to II-VI QDs, an aqueous CdTe QD solution was mixed under alkaline conditions with ethanol solution containing (3-mercaptopropyl)trimethoxysilane (MPS, a silane coupling agent) for the partial exchange of TGA ligands. A controlled amount of ethanol and MPS facilitated the formation of CdTe QD assemblies.

Step 2—Silica Layer Formation via the Stöber Method: A silica coating was developed on the assembled QDs using TEOS and aqueous ammonia via the Stöber method. The resulting silica nanoparticles were collected by centrifugation and re-dispersed in pure water.

4.2.1. Nanoparticle Morphology and PL Properties

The typical morphologies of silica nanoparticles prepared in Reference [61] are shown in Figure 7a–d. In the best-case scenario, PLQY was reduced slightly from its initial value (46%) to 40% upon encapsulation. Given the close packing of QDs within each nanoparticle, PL re-absorption occurs, similar to the behavior described in Section 3.2. This results in an overall red shift in the PL spectrum and a narrowed FWHM as shown in Figure 7e. By adjusting the MPS concentration in Step 1, the number of QDs per nanoparticle could be effectively controlled. Table 3 outlines the relationships between various parameters, including PL peak wavelength, PLQY, FWHM, nanoparticle size, and QD count.

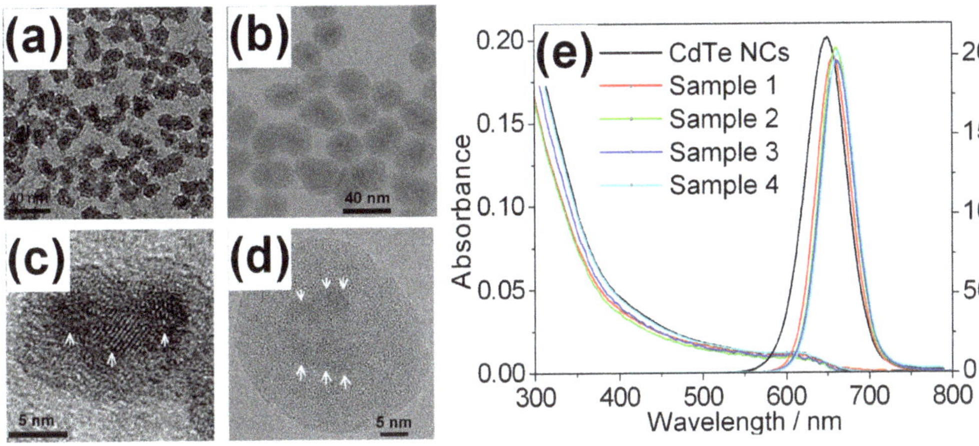

Figure 7. TEM images of luminescent silica nanoparticles. (**a**) Sample 2, (**b**) Sample 3, (**c**) and Sample 2 at a higher resolution, and (**d**) Sample 3 at a higher magnification. Three and six QDs (indicated by white arrows) are visible in (**c**) and (**d**), respectively. Well-developed lattice fringes are observed in CdTe QDs, confirming their structural integrity. (**e**) Absorption and PL spectra of luminescent silica nanoparticles, compared with those of initial CdTe QDs (nanocrystals, NCs). After encapsulation, PL peaks are red-shifted and a spectral narrowing is observed compared with original QDs [61].

Table 3. Properties of luminescent nanoparticles [61] [(a)].

Sample	PL Peak /nm	PLQY (%)	FWHM /nm	Mean Size /nm [(b)]	Average QDs No. in Each Nanoparticle [(b)]
1	658.4	34	49.2	12.5 ± 1.7	1.6 ± 0.7
2	660.8	40	48.0	15.9 ± 1.6	2.9 ± 0.9
3	662.6	31	48.4	29.1 ± 3.8	4.0 ± 1.6
4	663.8	32	47.2	35.9 ± 4.4	5.7 ± 2.3
CdTe [(a)]	650.2	46	58.0	3.9 ± 0.2	—

[(a)] Properties of initial CdTe QDs are shown for comparison. The molar ratios of MPS to QDs for Samples 1, 2, 3, and 4 during preparation were 176, 117, 57, and 29, respectively. [(b)] Estimated by TEM observation.

4.2.2. Hybrid Structure and Porous Properties

The silica nanoparticles exhibit a hybrid structure in the central part, primarily due to the combined presence of MPS and TEOS used in Step 1. The MPS concentration has a decisive role to control the number of QDs in the assembly. Higher MPS concentration results in fewer QDs in the central region after Step 2. This effect is quantitatively explained in the footnote of Table 3.

The porous nature of the silica nanoparticles thus prepared was evaluated by N_2 adsorption–desorption isotherms. When the silica layer thickness exceeded ~10 nm, the shielding ability improved due to a pore transition from cylindrical to ink-bottle-shaped pores. These structural behaviors differ significantly from those observed in bulk silica systems [62], where drying-induced shrinkage plays a key role in pore formation. Based on the insights into porous characteristics, a shell thickness exceeding 10 nm was prepared via the Stöber method, as described in the following section using hydrophobic QDs. The excellent shielding effects are detailed in Section 5.4.

The potential for biomedical applications was demonstrated by successfully conjugating biotinylated immunoglobulin to the prepared silica nanoparticles with a shell thickness of ca. 10 nm [61]. Quite similar results were obtained using aqueous InP-based QDs as well [63]. These suggest promising future directions for bio-tagging technologies utilizing QD-incorporated silica nanoparticles.

4.3. *Hydrophobic QDs Incorporated via Reverse Micelle Method*

Following the successful synthesis of highly emitting CdSe-based QDs [13,78], several researchers attempted to incorporate them into silica particles [64,65]. The most comprehensive study on encapsulation mechanisms was conducted by Meijerink's group in the Netherlands [66]. In their approach, CdSe-based QDs, typically octadecylamine-capped, were mixed with TEOS, surfactant, and aqueous ammonia in cyclohexane. This process rapidly exchanged the pristine alkyl-amine ligands, rendering the QDs hydrophilic at the expense of a significant PLQY reduction. The encapsulation process followed similar steps to those used for hydrophilic QDs, yet the final silica particles exhibited PLQY decay to below a few percent typically within one week.

A similar encapsulation approach was reported by another research group, involving oleic-acid coated CdSe-based QDs, instead of those coated with long-chain thiols. While the choice of the surfactant used in microemulsion system differed, the study similarly observed a time-dependent reduction in PLQY [67].

Single CdSe-based QDs were successfully encapsulated within individual silica particles, emphasizing surface silanization as a key factor in maintaining PL stability throughout the reverse micelle process [68]. For red-emitting QDs, the initial PLQY of 60% decreased to 35% post-encapsulation, while the PL spectral width (FWHM) narrowed from 28 nm to 24 nm. Impressively, the PLQY remained stable for months in aqueous conditions. Further investigations revealed that silanization conditions, particularly the molar ratio between QD and TEOS, play a crucial role in enhancing photostability [69].

4.4. *Hydrophobic QDs Incorporated via the Stöber Method*

An early study described the encapsulation of CdSe/ZnS QDs following ligand exchange with MPS. Researchers explored conditions for incorporating single QDs within individual silica particles [70], though PLQY values before and after encapsulation were not reported.

Around the same time, another study [71] demonstrated the formation of silica spheres incorporating CdS/ZnS QDs localized in the surface region. Initially, hydrophobic tri-n-octylphosphine oxide (TOPO) ligands on the QDs were replaced with two types of amino compounds, enhancing ethanol dispersibility. Similarly to the approach used for hydrophilic QDs, these vitreophilic QDs were then embedded within a silica layer (~100 nm thick) surrounding monodisperse silica microspheres (~550 nm in diameter). However, the resultant PLQY dropped from 38% to 13%. A limitation of this method is that the central silica core does not contribute to PL, reducing overall intensity.

Seemingly, the Stöber method appears to reduce the PLQY significantly because of the required ligand exchanges that facilitate the deposition of silica molecules onto the QDs. Additionally, lower alcohols commonly used to precipitate both hydrophilic and hydrophobic QDs often compromise their stability. Recognizing these challenges, the authors developed a new approach to integrate dozens of hydrophobic QDs into the central region of nanosized silica spheres while preserving the original PL properties. This approach is discussed in detail in the next section.

5. Silica Nanoparticles with Encapsulated Dozens of Hydrophobic CdSe-Based QDs

5.1. Preparation and Formation Mechanism

As discussed earlier [61], a silica layer thicker than 10 nm is crucial for effective shielding against the environment. However, encapsulating a single QD per silica particle results in an overall particle size of nearly 30 nm, which limits the high dispersion concentration required for applications. Consequently, smaller silica particles (several tens of nanometers in size) with dozens of embedded QDs have proven to be more suitable. Hydrophobic QDs synthesized at approximately 300 °C exhibit high PLQY and narrow FWHM, making them preferable for phosphor applications rather than hydrophilic QDs (as described in Section 4.2). Nonetheless, the synthetic method utilizing MPS as a size regulator remains a valuable approach. The encapsulation of CdSe-based QDs consists of three steps, as illustrated in Figure 8 [72,73].

Step 1—Surface Silanization: TEOS is added to a toluene solution containing hydrophobic QDs capped with oleic acid. Since analytical-grade toluene contains a small amount of water (~0.02 wt% [43]), a single alkoxy group among the four in TEOS gradually hydrolyzes, forming $(CH_3CH_2O)_3$-Si-OH, which replaces the pristine ligand molecule, a process termed surface silanization. Due to the slow ligand replacement, the partially hydrolyzed TEOS forms a well-ordered arrangement on the QD surface, ensuring effective passivation without changing the PLQY and FWHM of the initial QD. This step enhances the QD stability against lower-alcohol exposure.

Step 2—Formation of QD Assembly (Seed): The silanized QDs in toluene are mixed with hydrolyzed MPS in an aqueous ammonia and ethanol solution. Since the QDs in toluene are efficiently exposed to the aqueous phase because of abundant ethanol, they are hydrolyzed more, become hydrophilic, and finally are transferred into the aqueous phase. In this aqueous phase, QDs are condensed rapidly to form seed structures. Since MPS hydrolyzes slower than TEOS, MPS acts as a size regulator, similar to its role in hydrophilic QD encapsulation. A higher concentration of MPS results in smaller QD assemblies, with a lower QD count per particle.

Step 3—Silica Layer Formation via the Stöber Method: Following purification of the solution in Step 2, the Stöber method is applied to deposit a silica layer around the assembled QD structures, completing the encapsulation process.

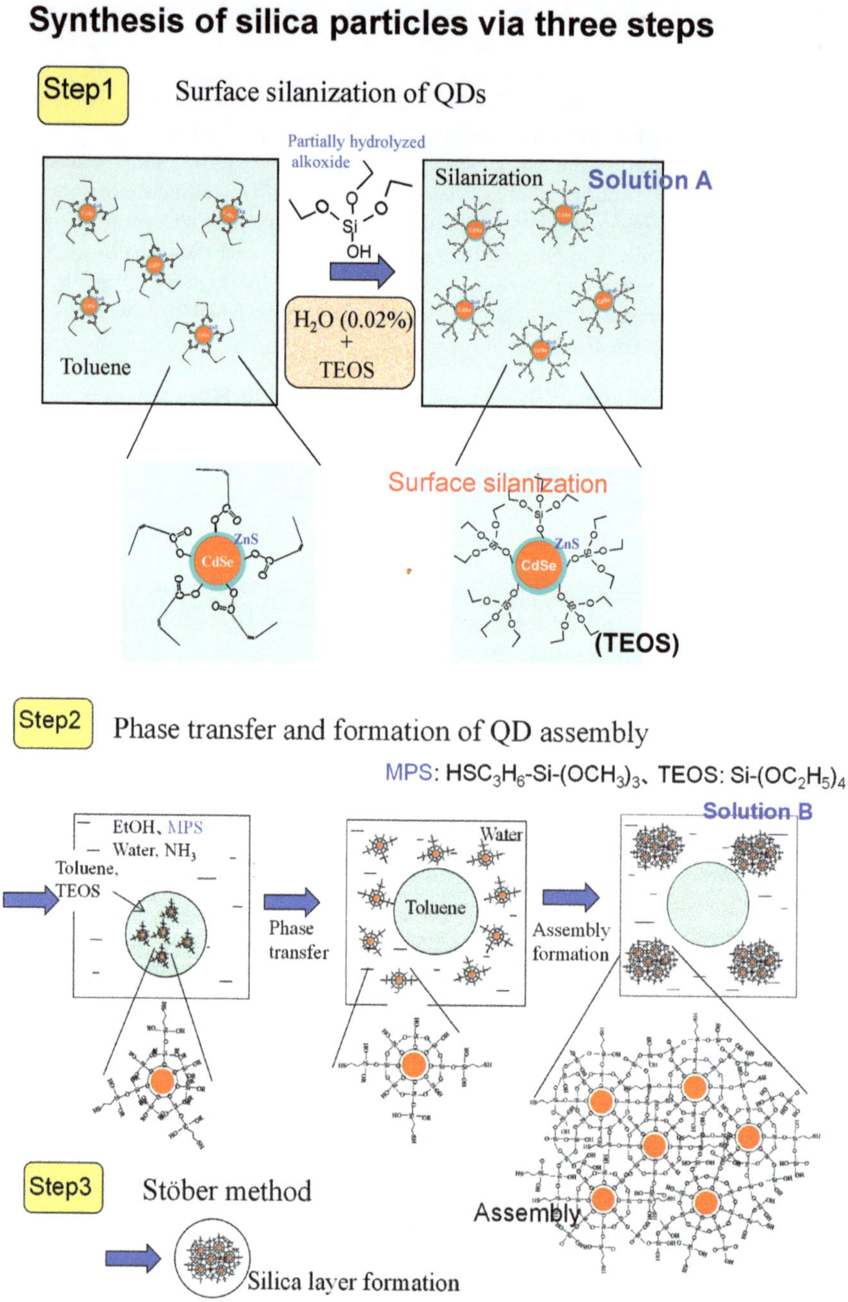

Figure 8. Schematic illustration of the process for assembling multiple CdSe/ZnS QDs into silica nanoparticles. The procedure consists of three key steps: (1) surface silanization, where QDs are modified to enhance compatibility with silica precursors; (2) phase transfer, facilitating QD transition from an organic to an aqueous phase; and (3) growth of the assembly, leading to encapsulation within silica matrices. Solution A contains silanized CdSe/ZnS QDs dispersed in toluene; Solution B consists of partially hydrolyzed MPS in ethanol, H_2O, and ammonium hydroxide (NH_4OH), essential for controlled nanoparticle formation [72].

5.2. Structural and Optical Properties

The morphology and PL spectra of the prepared silica nanoparticles are presented in Figure 9. The assembly size was successfully controlled by adjusting MPS concentration, as evidenced by the differences between Figure 9a and b. Additionally, Figure 9c confirms that PL spectral shape remains intact after the encapsulation. A three-dimensional imaging study [72] further reveals QD distribution within silica nanoparticles, distinctly visible as white dots. Initially, the synthesis focused solely on QDs emitting at ~620 nm, as shown in Figure 9c. To extend applicability, the preparation method was optimized for various PL wavelengths. It was determined that the QD concentration in Step 1, relative to the TEOS amount, must be optimized to achieve the maximum PLQY in silica nanoparticles. This optimization is critical because silanization efficiency in Step 1 depends on the total QD surface area in solution.

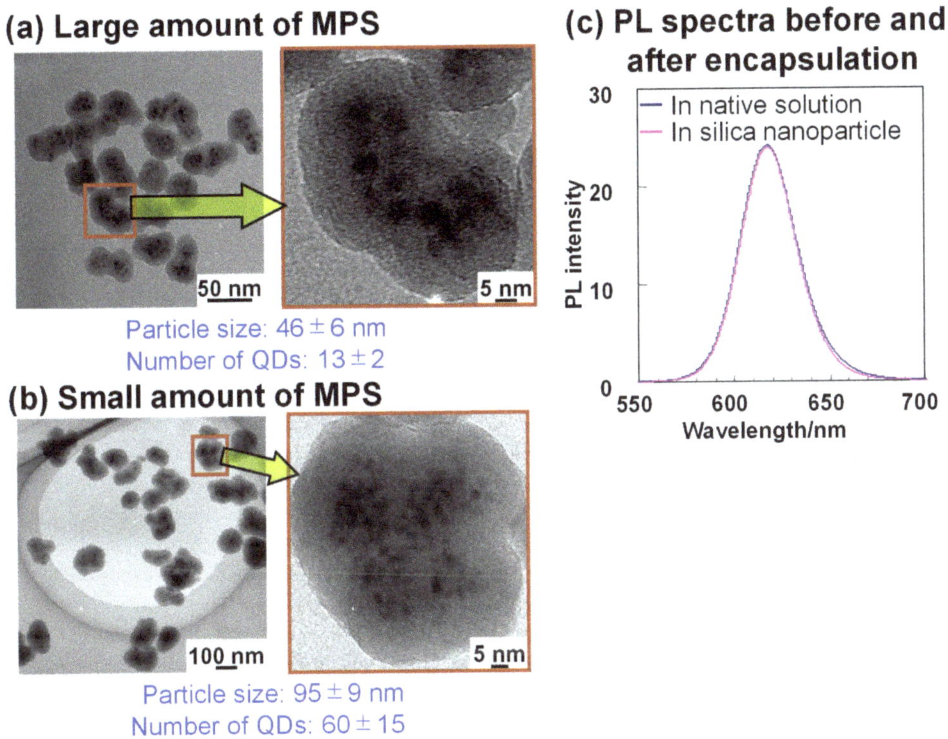

Figure 9. TEM images of luminescent silica nanoparticles displaying (**a**) particle 1 and (**b**) particle 2. The high-resolution images on the right provide a detailed visualization of the distribution of QDs in nanoparticles [72]. (**c**) Comparison of PL spectra of the colloidal solution and nanoparticle shown in (**a**).

The resultant PL spectra and PL images emitting initially at 550, 610, and 650 nm before and after the encapsulation are shown in Figure 10a. Notably, when smaller QDs are encapsulated, PL red shifting becomes more pronounced due to the additional influence of the quantum confinement effect associated with the smaller size of the QDs [74], despite similar reabsorption phenomena being observed and previously described in Figures 5b and 7e.

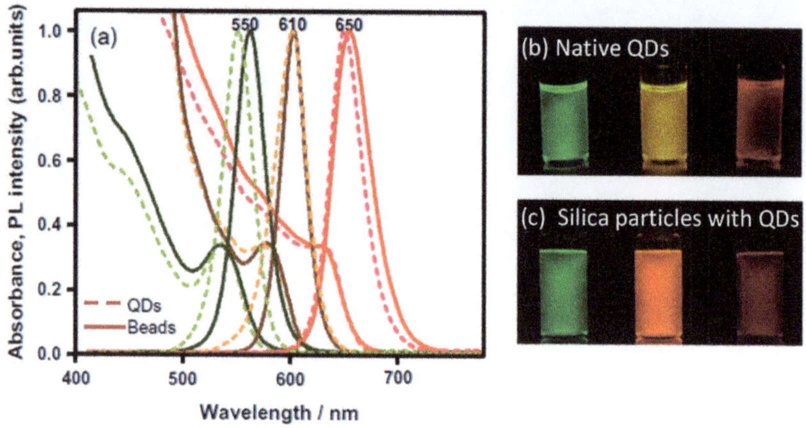

Figure 10. (**a**) Normalized absorption and PL spectra of CdSe/CdZnS QDs in toluene solution and in silica nanoparticles (beads). Numbers at the top of the graph indicate the PL peak wavelengths of QDs in toluene. (**b**) PL image of as-prepared QDs in toluene solution, and (**c**) PL image of QDs incorporated in silica particles, both excited by a UV lamp at 365 nm, illustrating emission characteristics before and after encapsulation [74].

5.3. PL Properties at the Single-Particle Level

Evaluations were conducted on the prepared silica nanoparticles to analyze their PL behavior at the single-particle level. Using a standard single-particle detection method, the PL intensity of two random pristine QDs exhibited distinct blinking, as illustrated in Figure 11a. However, in contrast, the PL intensity of two random silica nanoparticles displayed only fluctuations rather than blinking. Here, it is reasonable to consider that each individual QD within a single silica particle does not interact with others because the overall ensemble PL spectrum in Figure 9c remains unchanged after encapsulation. This means each QD in the particle exhibits the same blinking nature as pristine QDs before encapsulation. The PL originating from the ensemble of approximately 20 QDs per silica nanoparticle results in fluctuations within the measured temporal resolution of 200 ms. The apparent non-blinking behavior observed in Figure 11b is attributed to this exposure window, which masks the detail of individual blinking events. The variations likely reflect the cumulative blinking dynamics of the constituent QDs within the same nanoparticle. A crucial observation is that the PL intensity (vertical axis) from the silica nanoparticles in Figure 11 is approximately 15 times stronger than that of a single QD under similar conditions. The original fluorescence microscope recordings used for Figure 11, provide further evidence of this behavior for both pristine QDs and silica nanoparticles [72]. Notably, after continuous irradiation for 30 min, the silica nanoparticles exhibited significantly higher photostability compared to their pristine QD counterpart.

The observed substantially non-blinking behavior, coupled with strong PL from individual silica-encapsulated QD particles, is particularly advantageous for in vivo imaging applications [86]. This capability enables the simultaneous tracking of multiple biomolecular targets with enhanced spatial resolution and temporal resolution. Furthermore, silica surfaces offer well-established platforms for bioconjugation [87], facilitating targeted delivery and functionalization. Togher with the substantial suppression of heavy metal elusion discussed in the following subsection, these silica-encapsulated QDs represent highly promising fluorophores for biological applications.

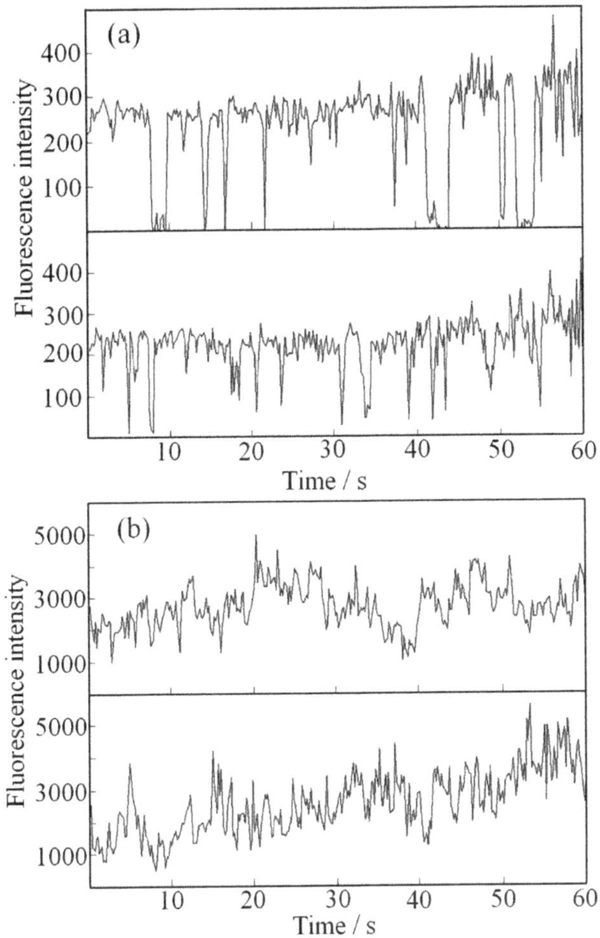

Figure 11. PL intensity trajectories of single QDs and silica particles, recorded with a temporal resolution of 200 ms over a duration of 60 s: (**a**) PL intensity of two random individual CdSe/ZnS QDs, exhibiting pronounced blinking behavior. (**b**) PL intensity of two random individual luminescent silica particles, demonstrating a non-blinking nature, with only minor fluctuations in intensity. Note that the average PL intensity of silica particles (**b**) was approximately 15 times higher than that of individual QDs (**a**) under comparable conditions, indicating enhanced emission in the silica matrix [72].

As outlined in the Introduction, several comprehensive review articles on biological applications of colloidal emitting QDs [4–6] have been published, following their initial development [13,78]. The most recent review [6] focuses on the pharmacokinetics and biodistribution of infrared-emitting QDs in murine models, demonstrating that surface modification and particle size critically influence distribution and excretion pathways. The silica-coating methodology discussed in this review is expected to significantly contribute to advancements in this promising research domain. Nonetheless, it is important to acknowledge that the extensive lead time and substantial cost associated with validation and regulatory approval continue to pose challenges for transitioning such technologies into clinical use for humans.

5.4. *Elusion of Heavy Metal Ions and Cytotoxicity of Silica Nanoparticles*

To assess the shielding performance of silica layers other than the photostability of PL intensity, two additional measurements were conducted to study the elusion of heavy metal ions (Cd^{2+}) into buffer solutions and cytotoxicity testing in live cell culture mediums [75].

For the elusion analysis, two additional silica nanoparticles were synthesized by modifying the Stöber method in Step 3. In Table 4, B represents the silica nanoparticle described in the previous Section 5.1, while C was prepared by performing the Stöber method at 40 °C instead of room temperature. To enhance biocompatibility for biological applications, the silica surface initially capped with hydroxyl (-OH) groups was modified with carboxyl (-COOH) groups for bioconjugation. This was achieved using carboxyethylsilanetriol sodium salt (CES) during the Stöber process [72], resulting in silica nanoparticle D. Additionally, a commercial polymer-coated aqueous QD (A10200, surface-modified with COOH, Thermo Fisher Scientific Inc., Waltham, MA, USA) was used as a control and labeled A.

Table 4. Concentration of dissolved Cd^{2+} from various QDs dispersed in HEPES solution.

No.	Sample	Concentration Ratio of Dissolved Cd^{2+} in HEPES Solution [(a)]	
A	Polymer-coated QDs (-COOH surface)	1	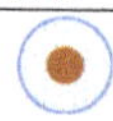
B	Silica nanoparticle with QDs (-OH surface, prepared at room temperature)	0.26	
C	Silica nanoparticle with QDs (-OH surface, prepared at 40 °C)	0.06	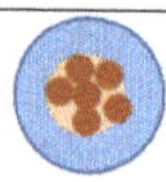
D	Silica nanoparticle with QDs (-COOH surface, prepared at room temperature)	0.009	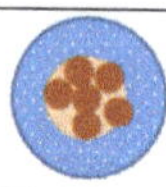

[(a)] The values are normalized using the measured concentration listed in Table 1 of Reference [75]. A schematic illustration depicting the development of each silica network is shown on the right-hand side.

Stock solutions of silica nanoparticles and commercial polymer-coated QDs listed in Table 4 were precipitated and redispersed in HEPES buffer solution (4-(2-hydroxyethyl)-1-piperazineethanesulfonic acid) at concentrations ranging from 20 to 50 nM. After 15 h, the solution was filtered through a centrifugal concentration filter. The concentration of Cd^{2+} in the obtained filtrate (in the ppb range) was quantified using inductively coupled plasma mass spectrometry (ICP-MS). Table 4 presents the normalized Cd^{2+} concentrations, adjusted to a standard 10 nM QD concentration, using A as the reference. The schematic diagrams on the right-hand side of the Table illustrate the morphology of QDs and silica matrix network development. As observed, silica matrices demonstrate superior shielding performance compared to polymer coatings, effectively reducing heavy metal ion elusion. Additionally, variations in Stöber method conditions significantly impact silica network formation, highlighting the importance of synthesis optimization. These insights contribute to understanding discrepancies in reported cytotoxicity data for silica shells following meta-analysis [88].

Cytotoxicity assessments complemented the heavy metal elusion study. The best silica nanoparticle (D) with a COOH-modified surface was compared to the control nanoparticle (A), which was also COOH-modified, using two widely recognized assays: MTT [3-(4,5-Dimethyl-2-thiazolyl)-2,5-diphenyltetrazolium Bromide] assay, which evaluates cell viability, and LDH (Lactate Dehydrogenase) assay, which assesses cell membrane damage.

Two widely studied human cell lines, lung carcinoma A549 cells and keratinocyte HaCaT cells, were used for the evaluation. Figure 12 demonstrates that D exhibited no cytotoxicity, whereas A induced significant toxicity in both cell types, as evidenced by MTT and LDH assay results.

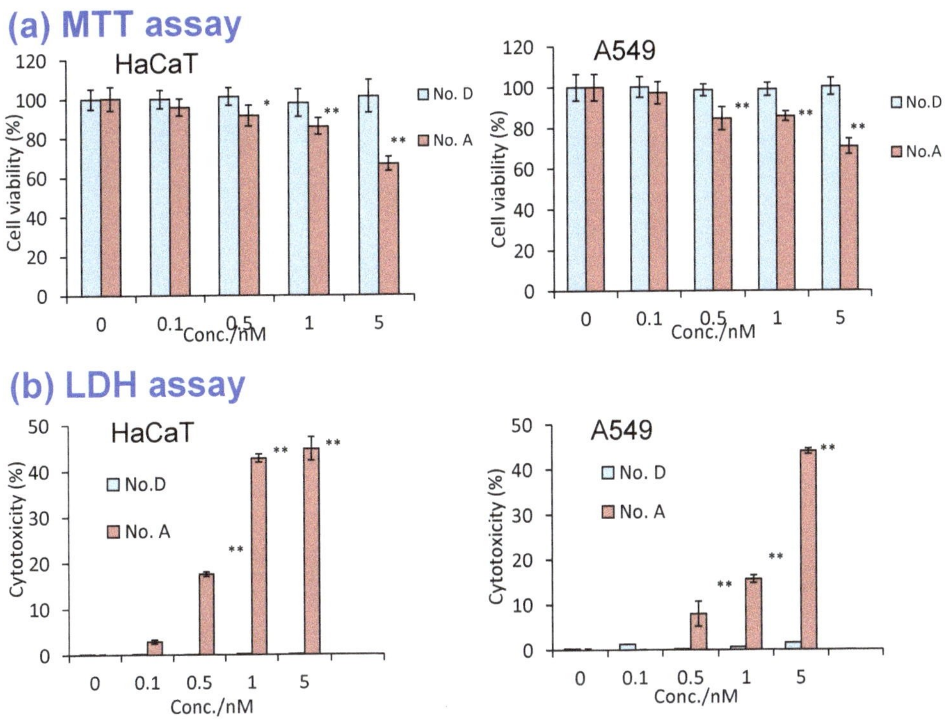

Figure 12. (**a**) MTT assay results for two human cell lines, HaCaT and A549, assessing cell viability based on mitochondrial activity. (**b**) LDH assay results for the same two cell lines, evaluating cell membrane damage. In this study, both mitochondrial function and membrane integrity were significantly affected only in cells exposed to polymer-coated QDs, whereas silica-encapsulated QDs demonstrated improved biocompatibility for both cell types [75]. Nos. A and D in the figure correspond to the samples A and D respectively shown in Table 4. All values are mean ± SD, * $p < 0.05$ vs. control, ** $p < 0.01$ vs. control. $p < 0.05$ is considered to indicate a statistically significant difference.

These cytotoxicity findings align with the heavy metal elusion trends shown in Table 4, reinforcing the correlation between silica shielding efficiency and biocompatibility. Moreover, silica layers synthesized via the Stöber method (Step 3) exhibit variations in shielding performance, suggesting that further optimization could be beneficial in developing robust silica matrices comparable to those widely utilized in semiconductor technology.

Expanding on Step 3, further investigations were conducted to assess how the preparation method, specifically the surface silanization conditions in Step 1, influences photostability [76]. Three types of nanoparticles, labeled ① to ③, were synthesized by varying only the initial ratio R of TEOS to QD, as shown in Figure 13, which directly impacted the QD density within the assembly and consequently the photostability of the silica particles. Morphological assessments were conducted using TEM and high-angle annular dark field scanning TEM (HAADF-STEM) (Figure 14). These images enabled precise quantification of QD count, assembly size, and silica shell thickness across all three nanoparticle types.

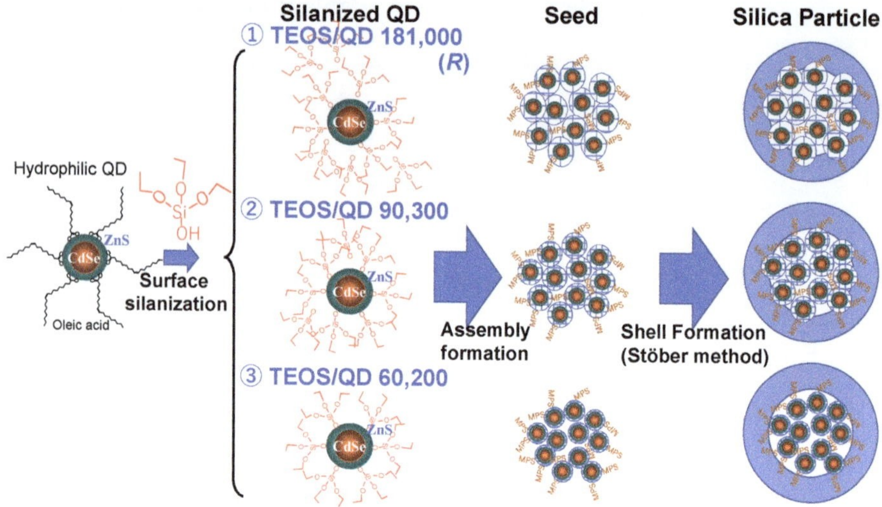

Figure 13. Reaction scheme illustrating the formation of three types (①–③) of silica particles, starting from hydrophobic QDs through surface silanization and assembly (seed) formation to shell (silica) formation. The initial ratio (*R*) of TEOS to QDs significantly influences the degree of condensation within the silica-glass network around the QDs in the particles. This dictates the difference in QD density and the photostability of silica particles [76].

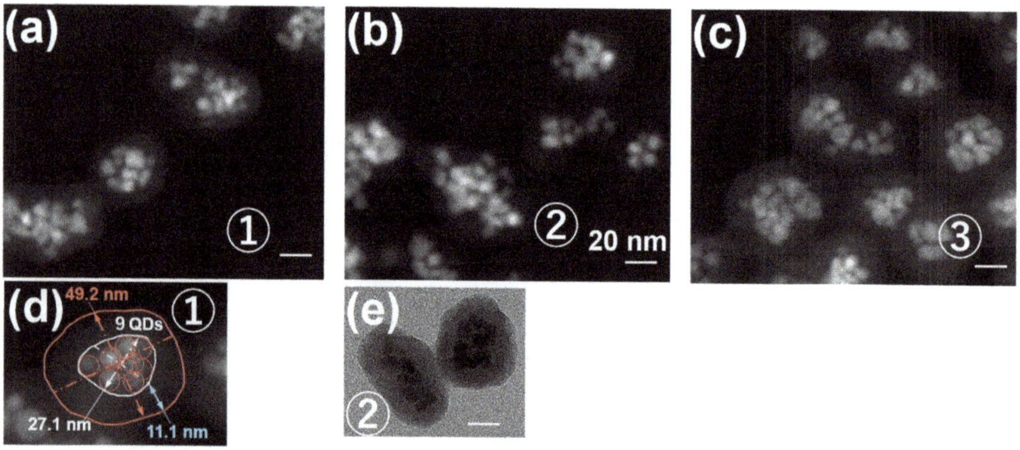

Figure 14. (**a**–**c**) HAADF images of three different types (①–③) of silica particles. The white scale bars in the images correspond to 20 nm. (**d**) Evaluation of the number of QDs present in type ① silica particles, including data on assembly (seed) size, overall particle dimensions, and shell thickness. The small red circles represent the outline of the QDs whereas the large red circle represents the outline of the entire nanoparticle. In addition, the white circle represents the outline of the seed. The particle size of nanoparticles and the thickness of silica shell are also shown. (**e**) Typical TEM image of type ② silica particles, corresponding to the HAADF image in (**b**) [76].

Figure 15 illustrates the correlation between assembly size and QD number, with corresponding linear fitting equations. Among the three samples, nanoparticle ③ exhibited the slope, indicating the highest packing density of QDs, which resulted from its unique silanization condition. To assess photostability, single-particle detection was applied to more than 20 individual nanoparticles from ①, ②, and ③. Figure 16a depicts the overall PL intensity decay over irradiation time for all three samples. A commercial polymer-coated aqueous CdSe-based QD (Qtracker™ 655, Thermo Fisher Scientific Inc. Massachusetts, USA), known for its exceptional durability, was selected as a control due to its emission wavelength being nearly identical to that of the silica nanoparticles. Figure 16b plots PL degradation time with error margins derived from Figure 16a, plotted against *R* (TEOS/QD ratio) and QD concentration during Step 1. The same Qtracker™ 655 data is displayed as a filled triangle on the vertical axis. These evaluations revealed that nanoparticle ③ exhibited up to an eight-fold increase in photostability compared to polymer-coated counterparts, achieved by optimizing the molar ratio *R* of alkoxide to QDs during silanization.

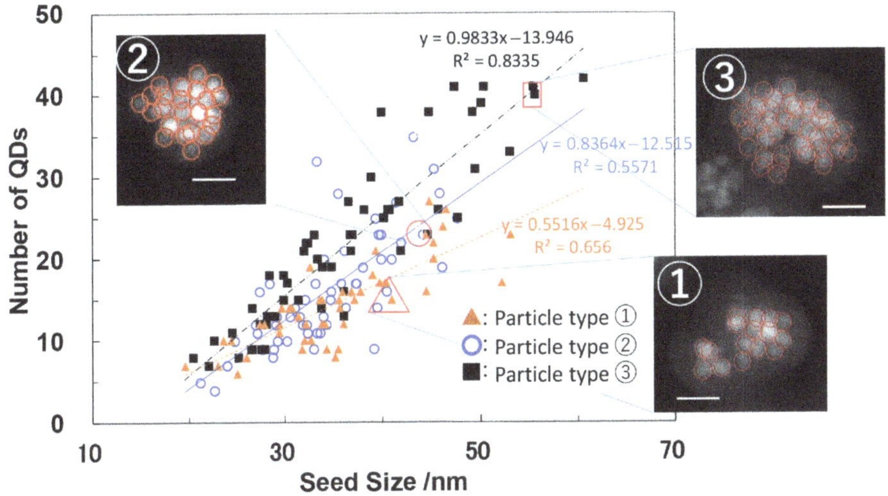

Figure 15. Relationship between the assembly (seed) size and the number of QDs for three types (①–③) of particles. The red circles in the images indicate the positions of QDs. The results of linear fitting are shown with corresponding equations and correlation coefficients. Three typical HAADF images used for counting the number of QDs in the particles are shown, with the data points marked by triangles, circles, and squares corresponding to types ①, ②, and ③, respectively. The white scale bars correspond to 20 nm [76].

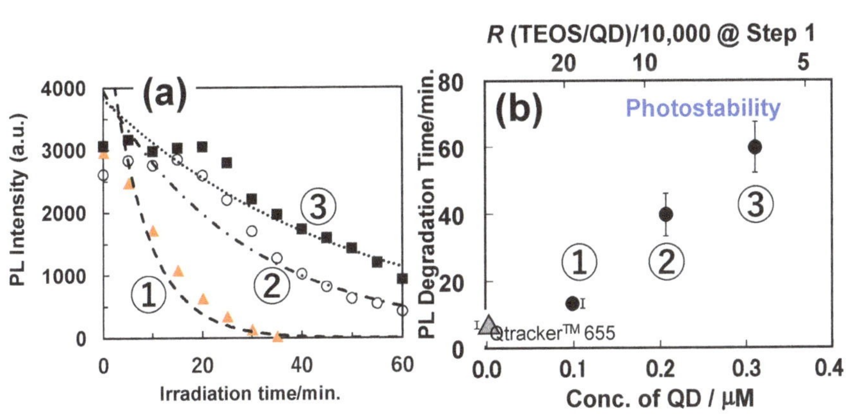

Figure 16. (**a**) PL intensity for three types (①–③) of nanoparticles against the irradiation time. Each data point was fitted using a single exponential decay curve. (**b**) PL degradation time with error bars for three types of nanoparticles. The result of commercial aqueous QD (Qtracker) is plotted on the vertical axis of the figure as a triangle for comparison [76].

The developed silica network formed in Step 1 near the QD surface remained structurally intact in Step 2, where hydrolyzed MPS was introduced to regulate assembly size. Ultimately, controlled silanization in Step 1 proved to be as significant as Step 3 in improving the photostability of silica nanoparticles.

6. Recent Progress and Applications

6.1. Adapting Earlier Approaches for Silica Nanoparticle Encapsulation with InP-Based QDs

As outlined in the background section, InP-based QDs stand out as the only environmentally benign binary system capable of emitting in the visible range. With this objective, the previously established silica nanoparticle encapsulation method, which was originally developed for hydrophobic CdSe-based QDs, was adapted for InP-based QDs. However, following this method, a significant decrease in PLQY was unexpectedly observed. To understand this discrepancy, three variations of InP-based QDs were synthesized, specifically InP/$(ZnSe)_n$/ZnS, with n = 4, 6, and 8 monolayers (MLs), each emitting at approximately 550 nm. These QDs were subsequently encapsulated within silica nanoparticles to investigate the influence of shell thickness and surface interactions on PLQY retention.

The TEM images and a schematic diagram of the prepared QDs are shown in Figure 17. Additionally, a HAADF-STEM image of one silica nanoparticle with embedded QDs (8 ML) is shown in Figure 18a. A single isolated QD in the silica matrix was selected for energy dispersive X-ray spectroscopy (EDX) analysis along the red horizontal line in Figure 18a, confirming in Figure 18b that the QD maintains its InP/ZnSe/ZnS structure. The findings suggest that QDs with a thicker intermediate ZnSe layer do not deteriorate significantly when encapsulated within silica nanoparticles.

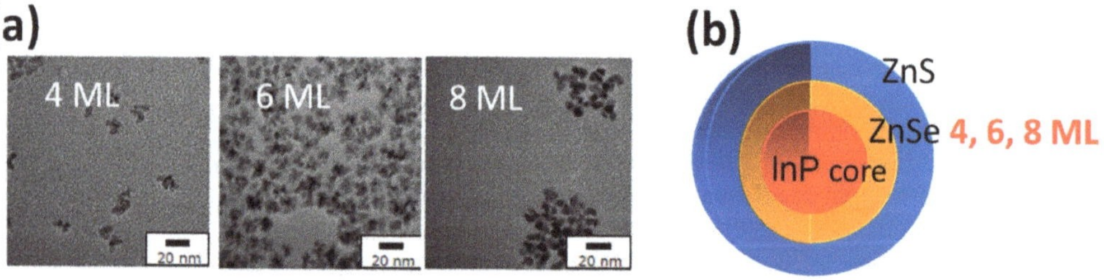

Figure 17. (**a**) TEM images of InP-based QDs synthesized with 4, 6, and 8 monolayers (ML) of ZnSe using the same core structure. (**b**) Schematic representation of the three types of InP/$(ZnSe)_n$/ZnS QDs (n = 4, 6, 8) used in this study [77].

Since the nanoparticle encapsulation process involves initial hydrophobic surface ligand exchange with partially hydrolyzed TEOS molecules, the QD surface conditions significantly influence the resulting PL properties. To further elucidate the underlying photophysical behavior, quantum mechanical simulations were carried out to calculate the electron distribution within the exciton for these three InP-based QDs. One example of a QD with an 8 ML intermediate ZnSe layer is shown in Figure 19. As depicted in the right-hand region of Figure 19, the electron spreads out from the QD structure into the surrounding environment. The relationship between ZnSe-shell thickness and electron spread is depicted by a line (a) in Figure 20, revealing a linear trend when plotted on a logarithmic scale. It is important to note that CdSe-based QDs, discussed in the previous section, typically feature shell thicknesses between 0.5 and 1.5 nm. Compared to CdSe-based QDs, InP-based QDs tend to exhibit greater electron dispersion due to their lighter effective electron mass and lower barrier height from the InP core to ZnSe/ZnS shells. As the ZnSe layer thickness increases, the amount of spread electron reduces, thereby better preserving the PLQY after encapsulation. On the other hand, the degree of electron spread outside the QD serves as a key factor in PLQY reduction for InP-based QDs. The simulations further indicate that the preferable structure for QDs includes an InP-core size greater than 2.2 nm with a ZnSe shell thickness of 2.5 nm or more, resulting in an electron spread outside the QD in the order of 10^{-5} or less, as shown in the figure [77].

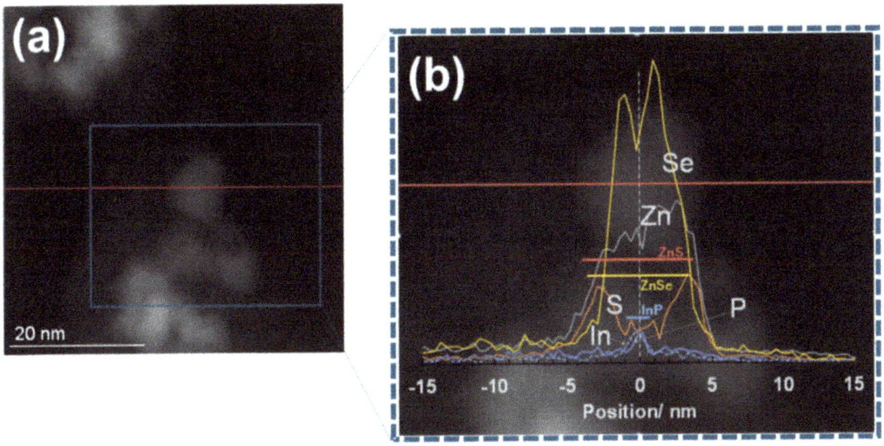

Figure 18. (**a**) Enlarged HAADF-STEM image of silica particles with encapsulated InP-based QDs (8 ML). The EDX analysis was performed along the red line across a single isolated QD. (**b**) EDX results are overlaid on the HAADF-STEM image. The horizontal bars (InP, ZnSe, ZnS) are drawn, corresponding to the preparation conditions used for QD synthesis and encapsulation [77]. The white vertical dashed line in (**b**) indicates the center line of the QD to be analyzed.

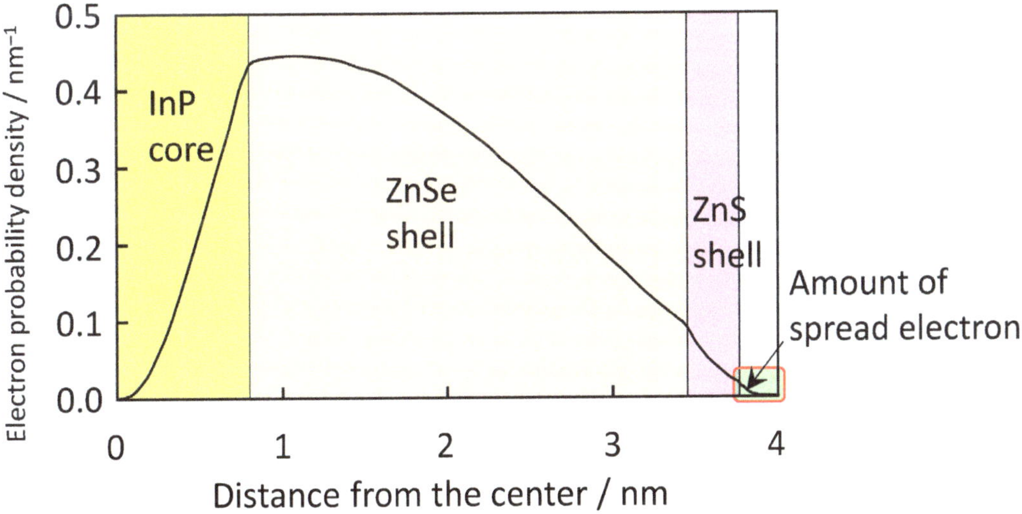

Figure 19. Radial probabilities for the presence of exciton electrons in QDs with 8 ML of ZnSe ($InP/(ZnSe)_8/ZnS$). The amount of spread electron (electron outside the QD) is shown on the right-hand side, illustrating the degree of confinement within the QD [77].

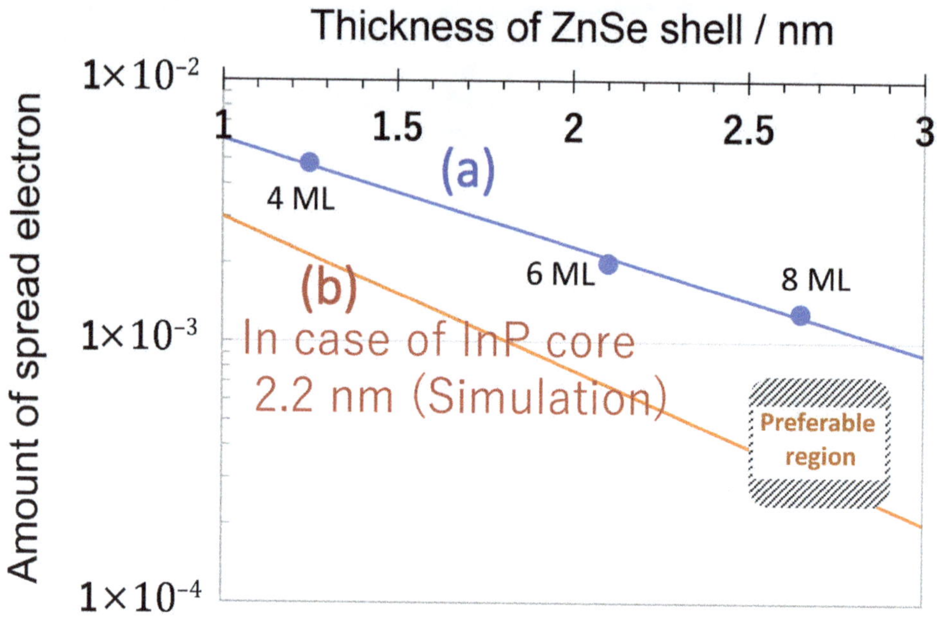

Figure 20. Relationship between the thickness of the ZnSe layer (InP/$(ZnSe)_{4,6,8}$/(ZnS)) and the amount of spread electron outside the QDs. (**a**) Three QDs investigated in this study. (**b**) Simulation for an InP core diameter of 2.2 nm as a function of ZnSe-layer thickness. The preferable structure region of the QD for silica encapsulation is indicated by diagonal lines [77].

6.2. Phosphors for Display Technologies and Mass Production Strategies

Figure 21 presents a schematic representation of the display panels currently available as commercial products. In the first-generation models, QDs were dispersed within a polymer layer (combining both green- and red-emitting QDs together with scattering materials such as titania) and sandwiched between barrier films, including glass films [2]. In subsequent generations, individual pixels were formed by separately assembled green- and red-emitting QDs, eliminating the need for liquid crystal while still requiring barrier films for stability. The robust silica particles are expected to replace free QDs, particularly those surrounded by polymers, thereby enhancing durability without necessitating additional stability measures during usage, as depicted at the bottom of Figure 21.

One of the main challenges in developing silica-encapsulated QDs was the lengthy preparation protocol, spanning steps 1 through 3 and requiring centrifugation at 50,000× *g* for separation. However, recent refinements have allowed the protocol to be streamlined into a single-pot process, reducing the necessary centrifuge speed to nearly half, significantly improving scalability for mass production. These silica particles are now being explored for broader applications beyond displays, offering a user-friendly phosphor with narrow spectral width, high brightness, and robust durability. Future investigations will focus on infrared emission using QDs composed of alternative heavy metals, particularly for biomedical applications [6] and infrared detection [89].

To drive commercialization, a start-up company [90] has been launched to leverage the findings presented in this article, particularly in optimizing phosphor materials for display technologies and expanding into infrared-emitting QDs. This effort entails detailed technical refinement and strategic negotiations to facilitate efficient manufacturing processes.

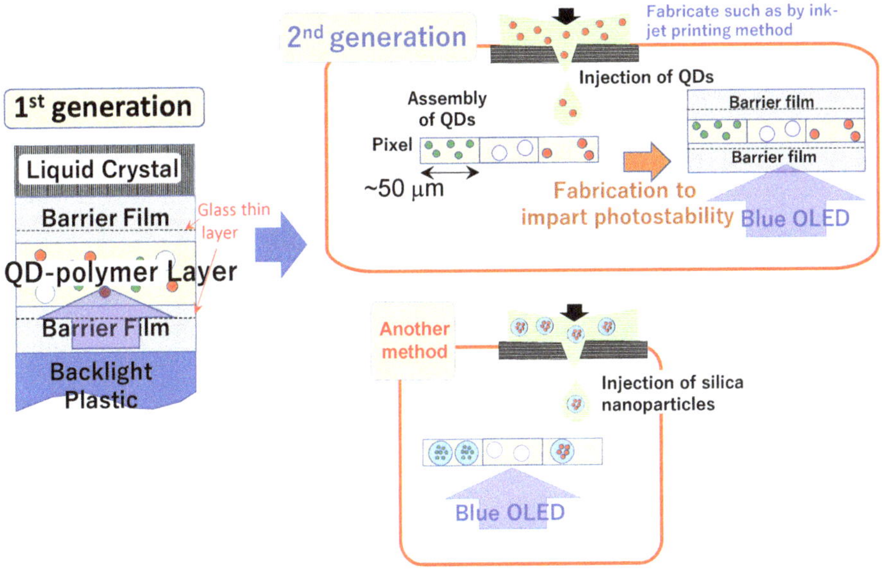

Figure 21. Schematic representation of display panel structures. In the first generation, the QDs are dispersed within a polymer layer and sandwiched between barrier films, including glass films. In the next generation, each pixel is formed by QD assembly. To enhance long-term stability, robust silica particles are proposed as a substitute for free QDs, reducing the need for additional stability measures during operation.

7. Concluding Remarks

In previous studies, the coating materials for QDs were compared, specifically polymer and silica [15]. Currently, more than five million QD displays are manufactured annually worldwide [3] using polymer coatings. However, due to their limited robustness, barrier films remain necessary. Ongoing research has actively addressed this challenge by fabricating silica nanoparticles encapsulating both hydrophobic and hydrophilic QDs. This dual encapsulation approach allows for broader material compatibility and facilitates comprehensive evaluation of photostability, dispersion uniformity, and biocompatibility across varied surface chemistries. The uses of silane coupling agents (initially APS and later MPS) have proved essential in controlling the overall morphology and maintaining the unique spectral properties of QDs.

Among these developments, silica nanoparticles with encapsulated dozens of hydrophobic CdSe-based QDs, synthesized through a three-step process using MPS (No. F in Table 2), demonstrated superior shielding effects. This was validated through PL measurements at the single-particle level, assessments of heavy metal ion (Cd^{2+}) elusion, and cytotoxicity tests. The preparation steps, particularly steps 1 and 3, were found to significantly affect the effectiveness of the shielding. Furthermore, when applying this encapsulation method to hydrophobic InP-based QDs (InP/ZnSe/ZnS), a considerable decrease in PLQY was observed. Experimental comparisons and quantum mechanical calculations revealed that maintaining a thick intermediate ZnSe layer (greater than 2.5 nm) was necessary to mitigate this deterioration. This requirement arises due to the lower effective mass of electrons and the shallow barrier height between the core and shell, distinguishing InP-based QDs from their CdSe-based counterparts.

Since emitting QDs cannot withstand prolonged exposure to high temperatures, sol–gel preparations were conducted at moderate temperatures, typically in the range of one hundred degrees Celsius. The findings from this research indicate that with careful optimization, silica layers can achieve an unprecedented level of shielding performance even at relatively low temperatures, presenting promising opportunities for future applications.

Author Contributions: N.M. conceived the central theme and structure of this review article and was responsible for writing most of the manuscript. C.L. contributed by enhancing the clarity of the writing, ensuring the accuracy of the scientific content, and assisting in formulating responses to reviewers' comments.

Funding: The works conducted by N. Murase reviewed in this article were financially supported mainly by Nano-glass Project (New Energy and Industrial Technology Development Organization, NEDO, Japan, 2002-2005), Core Research for Evolutional Science and Technology (CREST, Japan Science and Technology Agency, JST, 2006-2012), Creation and Support Program for Start-ups from Universities (JST, 2007-2009), Regional Innovation Creation Research and Development Project (Ministry of Economy, Trade and Industry, METI, 2010-2011), Supporting Industry Program (METI, 2010-2012), KAKENHI JP24550242 in Inorganic Materials (Japan Society for the Promotion of Science, JSPS, 2012-2015), Entrepreneurs Program A (NEP Type-A, NEDO, 2021), Entrepreneurs Program B (NEP Type-B, NEDO, 2022), and Post-Corona Startup Support Project (Hyogo prefecture, Japan, 2023).

Acknowledgments: We sincerely appreciate Ping Yang for her dedicated contributions during her time at AIST. We gratefully acknowledge Chie Hosokawa (Osaka Metropolitan University) for her insightful discussions and contributions to single-particle spectroscopy measurements conducted during her time at AIST. Much of the research outlined here was conducted in collaboration with N.M. while C.L. held a postdoctoral position at AIST. The establishment of QMT was driven by the goal of applying the technologies developed by N.M. at AIST. Among the funding agencies described above, NEDO has offered invaluable encouragement, supporting our efforts intermittently yet consistently for over two decades, dating back to the early 2000s. English proofreading of this article was performed by our former colleague Subramanian Tamil Selvan.

Conflicts of Interest: The authors declare no conflict of interest.

References

1. Chang, C.Y.; Sze, S.M. *Semiconductor Devices: Physics and Technology, International Student Version*, 3rd ed.; Wiley: Berlin, Germany, 2012.
2. Chen, J.; Hardev, V.; Yurek, J. Quantum-dot displays: Giving LCDs a competitive edge through color. *Inf. Disp.* **2013**, *29*, 12–17. [CrossRef]
3. Quantum Dot Display Market Size & Trends. Available online: https://www.grandviewresearch.com/industry-analysis/quantum-dot-qd-display-market (accessed on 21 June 2025).
4. Medintz, I.; Uyeda, H.; Goldman, E.; Mattoussi, H. Quantum dot bioconjugates for imaging, labelling and sensing. *Nat. Mater.* **2005**, *4*, 435–446. [CrossRef]
5. Kairdolf, B.; Smith, A.; Stokes, T.; Wang, M.; Young, A.; Nie, S. Semiconductor Quantum Dots for Bioimaging and Biodiagnostic Applications. *Annu. Rev. Anal. Chem.* **2013**, *6*, 143–162. [CrossRef]
6. Xu, G.; Zeng, S.; Zhang, B.; Swihart, M.; Yong, K.; Prasad, P. New Generation Cadmium-Free Quantum Dots for Biophotonics and Nanomedicine. *Chem. Rev.* **2016**, *116*, 12234–12327. [CrossRef]
7. What is the "Quantum Dots" That Won the Nobel Prize in Chemistry in 2023? Available online: https://www.aist.go.jp/aist_j/magazine/20231227.html (accessed on 21 June 2025).
8. Ekimov, A.I.; Onushchenko, A.A. Quantum size effect in the optical-spectra of semiconductor micro-crystals. *Sov. Phys. Semicond.* **1982**, *16*, 775–778.
9. Rossetti, R.; Nakahara, S.; Brus, L.E. Quantum size effects in the redox potentials, resonance Raman spectra, and electronic spectra of CdS crystallites in aqueous solution. *J. Chem. Phys.* **1983**, *79*, 1086–1088. [CrossRef]
10. Murase, N.; Gaponik, N.; Weller, H. Effect of chemical composition on luminescence of thiol-stabilized CdTe nanocrystals. *Nanoscale Res. Lett.* **2007**, *2*, 230–234. [CrossRef] [PubMed]
11. Tang, X.; Yang, J.; Li, S.; Liu, Z.; Hu, Z.; Hao, J.; Du, J.; Leng, Y.; Qin, H.; Lin, X.; et al. Single Halide Perovskite/Semiconductor Core/Shell Quantum Dots with Ultrastability and Nonblinking Properties. *Adv. Sci.* **2019**, *6*, 1900412. [CrossRef] [PubMed]

12. Gaponik, N.; Talapin, D.; Rogach, A.; Hoppe, K.; Shevchenko, E.; Kornowski, A.; Eychmüller, A.; Weller, H. Thiol-capping of CdTe nanocrystals: An alternative to organometallic synthetic routes. *J. Phys. Chem. B* **2002**, *106*, 7177–7185. [CrossRef]
13. Murray, C.B.; Norris, D.J.; Bawendi, M.G. Synthesis and characterization of nearly monodisperse CdE (E = sulfur, selenium, tellurium) semiconductor nanocrystallites. *J. Am. Chem. Soc.* **1993**, *115*, 8706–8715. [CrossRef]
14. Directive 2011/65/EU of the European Parliament and of the Council of 8 June 2011 on the Restriction of the Use of Certain Hazardous Substances in Electrical and Electronic Equipment (Recast) Text with EEA Relevance. Available online: https://eur-lex.europa.eu/eli/dir/2011/65/oj (accessed on 21 June 2025).
15. Murase, N. Quantum Dot-Core Silica Glass-Shell Nanomaterials: Synthesis, Characterization, and Potential Biomedical Applications. In *Nanotechnologies for the Life Sciences*; Wiley-VCH Verlag GmbH: Berlin, Germany, 2011; pp. 393–427.
16. Mićić, O.I.; Sprague, J.; Lu, Z.; Nozik, A.J. Highly efficient band-edge emission from InP quantum dots. *Appl. Phys. Lett.* **1996**, *68*, 3150–3152. [CrossRef]
17. Talapin, D.; Gaponik, N.; Borchert, H.; Rogach, A.; Haase, M.; Weller, H. Etching of colloidal InP nanocrystals with fluorides: Photochemical nature of the process resulting in high photoluminescence efficiency. *J. Phys. Chem. B* **2002**, *106*, 12659–12663. [CrossRef]
18. Preparation of Semiconductor Nanocrystals with High Photoluminescence Efficiency for Labeling in Biology. Available online: https://www.aist.go.jp/aist_e/list/latest_research/2009/20090325/20090325.html (accessed on 21 June 2025).
19. Li, Y.; Hou, X.; Dai, X.; Yao, Z.; Lv, L.; Jin, Y.; Peng, X. Stoichiometry-Controlled InP-Based Quantum Dots: Synthesis, Photoluminescence, and Electroluminescence. *J. Am. Chem. Soc.* **2019**, *141*, 6448–6452. [CrossRef] [PubMed]
20. Won, Y.; Cho, O.; Kim, T.; Chung, D.; Kim, T.; Chung, H.; Jang, H.; Lee, J.; Kim, D.; Jang, E. Highly efficient and stable InP/ZnSe/ZnS quantum dot light-emitting diodes. *Nature* **2019**, *575*, 634–638. [CrossRef]
21. Nakamura, H.; Kato, W.; Uehara, M.; Nose, K.; Omata, T.; Otsuka-Yao-Matsuo, S.; Miyazaki, M.; Maeda, H. Tunable photoluminescence wavelength of chalcopyrite $CuInS_2$-based semiconductor nanocrystals synthesized in a colloidal system. *Chem. Mater.* **2018**, *18*, 3330–3335. [CrossRef]
22. Uematsu, T.; Wajima, K.; Sharma, D.; Hirata, S.; Yamamoto, T.; Kameyama, T.; Vacha, M.; Torimoto, T.; Kuwabata, S. Narrow band-edge photoluminescence from $AgInS_2$ semiconductor nanoparticles by the formation of amorphous III-VI semiconductor shells. *NPG Asia Mater.* **2018**, *10*, 713–726. [CrossRef]
23. Uematsu, T.; Tepakidareekul, M.; Hirano, T.; Torimoto, T.; Kuwabata, S. Facile High-Yield Synthesis of Ag-In-Ga-S Quaternary Quantum Dots and Coating with Gallium Sulfide Shells for Narrow Band-Edge Emission. *Chem. Mater.* **2023**, *35*, 1094–1106. [CrossRef]
24. Huang, H.; Bodnarchuk, M.; Kershaw, S.; Kovalenko, M.; Rogach, A. Lead Halide Perovskite Nanocrystals in the Research Spotlight: Stability and Defect Tolerance. *ACS Energy Lett.* **2017**, *2*, 2071–2083. [CrossRef]
25. Zhang, Q.; Wang, R.; Feng, B.; Zhong, X.; Ostrikov, K. Photoluminescence mechanism of carbon dots: Triggering high-color-purity red fluorescence emission through edge amino protonation. *Nat. Commun.* **2021**, *12*, 6856. [CrossRef]
26. Saitow, K. Bright silicon quantum dot synthesis and LED design: Insights into size-ligand-property relationships from slow- and fast-band engineering. *Bull. Chem. Soc. Jpn.* **2024**, *97*, uoad002. [CrossRef]
27. Kim, J.; Shim, K.; Han, J.; Joo, J.; Heo, N.; Seff, K. Quantum Dots of $[Na_4Cs_6PbBr_4]^{8+}$, Water Stable in Zeolite X, Luminesce Sharply in the Green. *Adv. Mater.* **2020**, *32*, 2001868. [CrossRef]
28. Qu, L.; Miller, G. Passivated Nanoparticles. U.S. Patent 9,425,253B2, 23 September 2010.
29. Li, Z.; Yao, W.; Kong, L.; Zhao, Y.; Li, L. General Method for the Synthesis of Ultrastable Core/Shell Quantum Dots by Aluminum Doping. *J. Am. Chem. Soc.* **2015**, *137*, 12430–12433. [CrossRef]
30. Yin, B.; Sadtler, B.; Berezin, M.; Thimsen, E. Quantum dots protected from oxidative attack using alumina shells synthesized by atomic layer deposition. *Chem. Commun.* **2016**, *52*, 11127–11130. [CrossRef]
31. Zhang, Y.; Ni, H. Diffusion of H, C, and O Components in Silicate Melts. In *Diffusion in Minerals and Melts*; Zhang, Y., Cherniak, D.J., Eds.; De Gruyter Brill: Beijing, China, 2010; Volume 72, pp. 171–225.
32. Song, W.; Kim, J.; Yang, H. Silica-embedded quantum dots as downconverters of light-emitting diode and effect of silica on device operational stability. *Mater. Lett.* **2013**, *111*, 104–107. [CrossRef]
33. Lin, Z.; Fei, X.; Ma, Q.; Gao, X.; Su, X. $CuInS_2$ quantum dots@silica near-infrared fluorescent nanoprobe for cell imaging. *New J. Chem.* **2014**, *38*, 90–96. [CrossRef]
34. Wada, C.; Iso, Y.; Isobe, T.; Sasaki, H. Preparation and photoluminescence properties of yellow-emitting $CuInS_2$/ZnS quantum dots embedded in TMAS-derived silica. *RSC Adv.* **2017**, *7*, 7936–7943. [CrossRef]
35. Foda, M.; Huang, L.; Shao, F.; Han, H. Biocompatible and Highly Luminescent Near-Infrared $CuInS_2$/ZnS Quantum Dots Embedded Silica Beads for Cancer Cell Imaging. *ACS Appl. Mater. Interfaces* **2014**, *6*, 2011–2017. [CrossRef]

36. Kim, H.; Jang, H.; Kwon, B.; Suh, M.; Kim, Y.; Cheong, S.; Jeon, D. In Situ Synthesis of Thiol-Capped $CuInS_2$-ZnS Quantum Dots Embedded in Silica Powder by Sequential Ligand-Exchange and Silanization. *Electrochem. Solid State Lett.* **2012**, *15*, K16–K18. [CrossRef]
37. Yin, D.; Pei, L.; Liu, Z.; Yang, X.; Xiang, W.; Zhang, X. Synthesis, Characterization, and Photoluminescence on the Glass Doped with $AgInS_2$ Nanocrystals. *Adv. Condens. Matter Phys.* **2015**, *2015*, 141056. [CrossRef]
38. Yang, M.; Peng, H.; Zeng, F.; Teng, F.; Qu, Z.; Yang, D.; Wang, Y.; Chen, G.; Wang, D. In situ silica coating-directed synthesis of orthorhombic methylammonium lead bromide perovskite quantum dots with high stability. *J. Colloid Interface Sci.* **2018**, *509*, 32–38. [CrossRef]
39. Chen, J.; Wu, H.; Zhang, W.; Huang, Y.; Zheng, J.; Shao, J.; Chi, Y. Silica coated perovskite quantum dots with single-particle level for dual-mode and smartphone-assisted detection of glutathione. *Sens. Actuator B-Chem.* **2025**, *432*, 137489. [CrossRef]
40. Kim, J.; Kim, B.; Kim, M.; Jang, W.; Wang, D. One-step formation of core/shell structure based on hydrophobic silane ligands for enhanced luminescent perovskite quantum dots. *J. Alloys Compd.* **2021**, *886*, 161347. [CrossRef]
41. Liu, Z.; Hu, Z.; Shi, T.; Du, J.; Yang, J.; Zhang, Z.; Tang, X.; Leng, Y. Stable and enhanced frequency up-converted lasing from CsPbBr3 quantum dots embedded in silica sphere. *Opt. Express* **2019**, *27*, 9459–9466. [CrossRef]
42. Zhong, Q.; Cao, M.; Hu, H.; Yang, D.; Chen, M.; Li, P.; Wu, L.; Zhang, Q. One-Pot Synthesis of Highly Stable $CsPbBr_3@SiO_2$ Core-Shell Nanoparticles. *ACS Nano* **2018**, *12*, 8579–8587. [CrossRef] [PubMed]
43. Huang, S.; Li, Z.; Kong, L.; Zhu, N.; Shan, A.; Li, L. Enhancing the Stability of $CH_3NH_3PbBr_3$ Quantum Dots by Embedding in Silica Spheres Derived from Tetramethyl Orthosilicate in "Waterless" Toluene. *J. Am. Chem. Soc.* **2016**, *138*, 5749–5752. [CrossRef]
44. Ye, L.; Hong, S.; Lu, C.; Zhao, Q. Stable random laser of perovskite quantum dots based on SiO_2-QDs-SiO_2 composite nanostructure. *J. Lumin.* **2025**, *277*, 120946. [CrossRef]
45. Hai, O.; Li, T.; Li, J.; Qin, B.; Ren, Q.; Wu, X.; Wu, Y. Preparation and properties of highly stabilized $CsPbX_3$ (X = Cl, Br, I) by liquid phase molten salt method. *Opt. Mater.* **2025**, *161*, 116824. [CrossRef]
46. Wang, H.; Lin, S.; Tang, A.; Singh, B.; Tong, H.; Chen, C.; Lee, Y.; Tsai, T.; Liu, R. Mesoporous Silica Particles Integrated with All-Inorganic $CsPbBr_3$ Perovskite Quantum-Dot Nanocomposites (MP-PQDs) with High Stability and Wide Color Gamut Used for Backlight Display. *Angew. Chem.-Int. Edit.* **2016**, *55*, 7924–7929. [CrossRef]
47. Di, X.; Jiang, J.; Hu, Z.; Zhou, L.; Li, P.; Liu, S.; Xiang, W.; Liang, X. Stable and brightly luminescent all-inorganic cesium lead halide perovskite quantum dots coated with mesoporous silica for warm WLED. *Dyes Pigment.* **2017**, *146*, 361–367. [CrossRef]
48. Di, X.; Hu, Z.; Jiang, J.; He, M.; Zhou, L.; Xiang, W.; Liang, X. Use of long-term stable $CsPbBr_3$ perovskite quantum dots in phospho-silicate glass for highly efficient white LEDs. *Chem. Commun.* **2017**, *53*, 11068–11071. [CrossRef]
49. Ekimov, A. Growth and optical properties of semiconductor nanocrystals in a glass matrix. *J. Lumin.* **1996**, *70*, 1–20. [CrossRef]
50. Murase, N.; Yazawa, T. Partially Reduced Cuprous Oxide Nanoparticles Formed in Porous Glass Reaction Fields. *J. Am. Ceram. Soc.* **2001**, *84*, 2269–2273. [CrossRef]
51. Selvan, S.; Li, C.; Ando, M.; Murase, N. Formation of luminescent CdTe-silica nanoparticles through an inverse microemulsion technique. *Chem. Lett.* **2004**, *33*, 434–435. [CrossRef]
52. Yang, P.; Ando, M.; Murase, N. Encapsulation of emitting CdTe QDs within silica beads to retain initial photoluminescence efficiency. *J. Colloid Interface Sci.* **2007**, *316*, 420–427. [CrossRef]
53. Teng, F.; Tian, Z.; Xiong, G.; Xu, Z. Preparation of CdS-SiO_2 core-shell particles and hollow SiO_2 spheres ranging from nanometers to microns in the nonionic reverse microemulsions. *Catal. Today* **2004**, *93–95*, 651–657. [CrossRef]
54. Selvan, S.; Bullen, C.; Ashokkumar, M.; Mulvaney, P. Synthesis of tunable, highly luminescent QD-glasses through sol-gel processing. *Adv. Mater.* **2001**, *13*, 985–988. [CrossRef]
55. Li, C.; Murase, N. Synthesis of highly luminescent glasses incorporating CdTe nanocrystals through sol-gel processing. *Langmuir* **2004**, *20*, 1–4. [CrossRef] [PubMed]
56. Kim, Y.; Lee, H.; Kang, S.; Bae, B. Two-Step-Enhanced Stability of Quantum Dots via Silica and Siloxane Encapsulation for the Long-Term Operation of Light Emitting Diodes. *ACS Appl. Mater. Interfaces* **2019**, *11*, 22801–22808. [CrossRef]
57. Watanabe, T.; Wada, C.; Iso, Y.; Isobe, T.; Sasaki, H. Preparation of Photostable Fluorescent InP/ZnS Quantum Dots Embedded in TMAS-Derived Silica. *ECS J. Solid State Sci. Technol.* **2017**, *6*, R75–R80. [CrossRef]
58. Watanabe, T.; Iso, Y.; Isobe, T.; Sasaki, H. Photoluminescence color stability of green-emitting InP/ZnS core/shell quantum dots embedded in silica prepared via hydrophobic routes. *RSC Adv.* **2018**, *8*, 25526–25533. [CrossRef]
59. Yang, P.; Li, C.; Murase, N. Highly photoluminescent multilayer QD-glass films prepared by LbL self-assembly. *Langmuir* **2005**, *21*, 8913–8917. [CrossRef]
60. Rogach, A.; Nagesha, D.; Ostrander, J.; Giersig, M.; Kotov, N. "Raisin bun"-type composite spheres of silica and semiconductor nanocrystals. *Chem. Mater.* **2000**, *12*, 2676–2685. [CrossRef]
61. Yang, P.; Murase, N. Size-Tunable Highly Luminescent SiO_2 Particles Impregnated with Number-Adjusted CdTe Nanocrystals. *ChemPhysChem* **2010**, *11*, 815–821. [CrossRef]

62. Brinker, C.J.; Scherer, G.W. Theory of Deformation and Flow in Gels. In *Sol-Gel Science*, 2nd ed.; Academic Press: Pittsburgh, PA, USA, 1990; pp. 406–451.
63. Li, C.; Hosokawa, C.; Suzuki, M.; Taguchi, T.; Murase, N. Preparation and biomedical applications of bright robust silica nanocapsules with multiple incorporated InP/ZnS quantum dots. *New J. Chem.* **2018**, *42*, 18951–18960. [CrossRef]
64. Darbandi, M.; Thomann, R.; Nann, T. Single quantum dots in silica spheres by microemulsion synthesis. *Chem. Mater.* **2005**, *17*, 5720–5725. [CrossRef]
65. Selvan, S.; Tan, T.; Ying, J. Robust, non-cytotoxic, silica-coated CdSe quantum dots with efficient photoluminescence. *Adv. Mater.* **2005**, *17*, 1620–1625. [CrossRef]
66. Koole, R.; van Schooneveld, M.; Hilhorst, J.; Donegá, C.; Hart, D.; van Blaaderen, A.; Vanmaekelbergh, D.; Meijerink, A. On the incorporation mechanism of hydrophobic quantum dots in silica spheres by a reverse microemulsion method. *Chem. Mater.* **2008**, *20*, 2503–2512. [CrossRef]
67. Drozd, D.; Zhang, H.; Goryacheva, I.; De Saeger, S.; Beloglazova, N. Silanization of quantum dots: Challenges and perspectives. *Talanta* **2019**, *205*, 120164. [CrossRef]
68. Yang, P.; Ando, M.; Murase, N. Highly Luminescent CdSe/$Cd_xZn_{1-x}S$ Quantum Dots Coated with Thickness-Controlled SiO_2 Shell through Silanization. *Langmuir* **2011**, *27*, 9535–9540. [CrossRef] [PubMed]
69. Murase, N.; Li, C.; Gunshi, T. Improved photostability of silica bead impregnated with CdSe-based quantum dots prepared through proper surface silanization. *Colloid Surf. A* **2020**, *600*, 124811. [CrossRef]
70. Nann, T.; Mulvaney, P. Single quantum dots in spherical silica particles. *Angew. Chem.-Int. Edit.* **2004**, *43*, 5393–5396. [CrossRef]
71. Chan, Y.; Zimmer, J.; Stroh, M.; Steckel, J.; Jain, R.; Bawendi, M. Incorporation of luminescent nanocrystals into monodisperse core-shell silica microspheres. *Adv. Mater.* **2004**, *16*, 2092–2097. [CrossRef]
72. Yang, P.; Murase, N.; Suzuki, M.; Hosokawa, C.; Kawasaki, K.; Kato, T.; Taguchi, T. Bright, non-blinking, and less-cytotoxic SiO_2 beads with multiple CdSe/ZnS nanocrystals. *Chem. Commun.* **2010**, *46*, 4595–4597. [CrossRef] [PubMed]
73. Fabrication of Photoluminescent Small Glass Capsules Containing Quantum Dots at High Concentration. Available online: https://www.aist.go.jp/aist_e/list/latest_research/2010/20100723/20100723.html (accessed on 21 June 2025).
74. Li, C.; Murase, N. Formation mechanism of highly luminescent silica capsules incorporating multiple hydrophobic quantum dots with various emission wavelengths. *J. Colloid Interface Sci.* **2013**, *411*, 82–91. [CrossRef]
75. Murase, N.; Horie, M.; Sawai, T.; Kawasaki, K. Silica layer-dependent leakage of cadmium from CdSe/ZnS quantum dots and comparison of cytotoxicity with polymer-coated analogues. *J. Nanopart. Res.* **2019**, *21*, 10. [CrossRef]
76. Murase, N.; Hosokawa, C.; Li, C.; Onishi, M.; Sawai, T. Drastic photostability improvement of silica particles impregnated with multiple emitting CdSe quantum dots prepared through efficient surface silanization. *Colloid Surf. A* **2022**, *654*, 130084. [CrossRef]
77. Murase, N.; Sawai, T.; Mori, R.; Inada, K.; Eguchi, D.; Tamai, N. Preparation of green-emitting InP-based quantum dots with controlled shell thickness and their photoluminescence quantum yield upon silica encapsulation. *J. Nanopart. Res.* **2025**, *27*, 75. [CrossRef]
78. Dabbousi, B.O.; Rodriguez-Viejo, J.; Mikulec, F.V.; Heine, J.R.; Mattoussi, H.; Ober, R.; Jensen, K.F.; Bawendi, M.G. (CdSe)ZnS Core–Shell Quantum Dots: Synthesis and Characterization of a Size Series of Highly Luminescent Nanocrystallites. *J. Phys. Chem. B* **1997**, *101*, 9463–9475. [CrossRef]
79. Lee, J.; Sundar, V.; Heine, J.; Bawendi, M.; Jensen, K. Full color emission from II-VI semiconductor quantum dot-polymer composites. *Adv. Mater.* **2000**, *12*, 1102–1105. [CrossRef]
80. Talapin, D.; Rogach, A.; Haase, M.; Weller, H. Evolution of an ensemble of nanoparticles in a colloidal solution: Theoretical study. *J. Phys. Chem. B* **2001**, *105*, 12278–12285. [CrossRef]
81. Yu, W.; Qu, L.; Guo, W.; Peng, X. Experimental determination of the extinction coefficient of CdTe, CdSe, and CdS nanocrystals. *Chem. Mater.* **2003**, *15*, 2854–2860. [CrossRef]
82. Han, J.; Liu, X.; Tan, J.; He, Z.; Li, C. Facile Method of Preparing Highly Luminescent Silica Gel Glass Incorporating Hydrophobic Semiconductor Quantum Dots. *Chem. Lett.* **2015**, *44*, 1434–1436. [CrossRef]
83. Liu, X.; Han, J.; Wu, W.; Shi, Q.; Li, W.; Li, C. Highly Luminescent Glass Films Incorporating Hydrophobic Quantum Dots Prepared by Layer-by-layer Self-assembly Method. *Chem. Lett.* **2016**, *45*, 10–12. [CrossRef]
84. Gerion, D.; Pinaud, F.; Williams, S.; Parak, W.; Zanchet, D.; Weiss, S.; Alivisatos, A. Synthesis and properties of biocompatible water-soluble silica-coated CdSe/ZnS semiconductor quantum dots. *J. Phys. Chem. B* **2001**, *105*, 8861–8871. [CrossRef]
85. Jana, N.; Earhart, C.; Ying, J. Synthesis of water-soluble and functionalized nanoparticles by silica coating. *Chem. Mater.* **2007**, *19*, 5074–5082. [CrossRef]
86. In Vivo Nano-Imaging of Movement of Molecules in Mice. Available online: https://www.jst.go.jp/kisoken/crest/en/area02/3-02.html (accessed on 21 June 2025).
87. Hermanson, G.T. *Bioconjugate Techniques*, 3rd ed.; Elsevier: Waltham, MA, USA, 2013; p. 02451.
88. Oh, E.; Liu, R.; Nel, A.; Gemill, K.; Bilal, M.; Cohen, Y.; Medintz, I. Meta-analysis of cellular toxicity for cadmium-containing quantum dots. *Nat. Nanotechnol.* **2016**, *11*, 479–486. [CrossRef] [PubMed]

89. Vafaie, M.; Fan, J.Z.; Najarian, A.M.; Ouellette, O.; Sagar, L.K.; Bertens, K.; Sun, B.; Arquer, F.P.G.; Sargent, E.H. Colloidal quantum dot photodetectors with 10-ns response time and 80% quantum efficiency at 1,550 nm. *Matter* **2021**, *4*, 1042–1053. [CrossRef]
90. Quantum Materials Technology. Available online: https://www.quantum-mater-tech.com/ (accessed on 21 June 2025).

Disclaimer/Publisher's Note: The statements, opinions and data contained in all publications are solely those of the individual author(s) and contributor(s) and not of MDPI and/or the editor(s). MDPI and/or the editor(s) disclaim responsibility for any injury to people or property resulting from any ideas, methods, instructions or products referred to in the content.

molecules

MDPI

Review

AI/Machine Learning and Sol-Gel Derived Hybrid Materials: A Winning Coupling

Aurelio Bifulco [1] and Giulio Malucelli [2,3,*]

1 Department of Chemical, Materials and Industrial Production Engineering (DICMaPI), University of Naples Federico II, Piazzale Tecchio 80, 80125 Napoli, Italy; aurelio.bifulco@unina.it
2 Department of Applied Science and Technology, Politecnico di Torino, Viale Teresa Michel 5, 15121 Alessandria, Italy
3 Consorzio Interuniversitario Nazionale per la Scienza e Tecnologia dei Materiali (INSTM) Unit, Via G. Giusti 9, 50121 Florence, Italy
* Correspondence: giulio.malucelli@polito.it; Tel.: +39-013-122-9369

Abstract

Experimental research in the field of science and technology of polymeric materials and their hybrid organic-inorganic systems has been and will continue to be based on the execution of tests to establish robust structure-morphology-property-processing correlations. Although absolutely necessary, these tests are often time-consuming and require specific efforts; sometimes, they must be repeated to achieve a certain reproducibility and reliability. In this context, the introduction of methods like the Design of Experiments (DoEs) has made it possible to drastically reduce the number of experimental tests required for a complete characterization of a material system. However, this does not seem enough. Indeed, further improvements are being observed thanks to the introduction of a very recent approach based on the use of artificial intelligence (AI) through the exploitation of a "machine learning (ML)" strategy: this way, it is possible to "teach" AI how to use literature data already available (and even incomplete) for material systems similar to the one being explored to predict key parameters of this latter, minimizing the error while maximizing the reliability. This work aims to provide an overview of the current, new (and up-to-date) use of AI/ML strategies in the field of sol-gel-derived hybrid materials.

Keywords: sol-gel method; hybrid materials; artificial intelligence; machine learning; decision trees; neural networks

1. Introduction

The scientific literature is rich with several interesting examples regarding the synthesis and characterization of sol-gel-derived hybrid organic-inorganic materials: indeed, these systems can be quite easily obtained by exploiting the occurrence of hydrolysis and condensation reactions of suitable precursors, possibly in the presence of coupling agents [1–4]. Additionally, the wide availability of the latter, together with the possibility of designing formulations with diverse types of monomers (epoxy, (meth)acrylic, vinyl ester, . . .), allows for obtaining hybrid materials that, at least potentially, are suitable for different applications in advanced sectors [5–13].

Despite their robustness and feasibility, the optimization of sol-gel-derived hybrid organic-inorganic systems requires the design, development, and implementation of testing activities aimed at identifying the best performing recipes and the best experimental parameters as well. Indeed, any changes in the starting formulations (chemical structure

 https://doi.org/10.3390/molecules30143043

and reactivity of selected monomers and precursors, monomers to precursors ratio, type and loading of coupling agents, among others) and in the experimental conditions (pH, temperature, reaction time) may significantly affect the overall sol-gel process and, subsequently, the properties of the final hybrid materials [14–17]. However, this optimization is often time-consuming and requires specific efforts, also considering that experimental testing may necessitate some replicas to achieve acceptable reproducibility and reliability.

Notwithstanding that the experimental activities carried out on any material systems are remarkably necessary, an important reduction in the number of the experimental tests needed for a complete characterization has been achieved by proposing such methods as the Design of Experiments (DoEs) [18]. Briefly, DoE represents a branch of applied statistics, which encompasses planning, running, and analyzing controlled tests to evaluate factors affecting the value of one or more parameters. It is, therefore, a powerful tool for collecting and analyzing data in a variety of experimental conditions; besides, it accounts for a reduction in the experimental testing efforts, hence simplifying the experimental activities to be performed.

This methodology has also been successfully applied in the case of sol-gel-derived hybrid materials.

In particular, Mahmoodian and co-workers [19] synthesized organic–inorganic hybrid systems based on a modified Bisphenol A diglycidyl methacrylate (bearing two ethoxysilane reactive groups), tetraethylorthosilicate (TEOS), and triethylene glycol dimethacrylate, suitable for orthodontic purposes (i.e., dental resins). The DoE approach based on the Taguchi orthogonal method was exploited to find the best experimental conditions for achieving the highest gel content without complete gelation of the resin mixture.

Najafabadi et al. [20] exploited the DoE methodology based on Taguchi orthogonal design to investigate and optimize compositional and process parameters of hybrid organic-inorganic coatings, sol-gel-derived from TEOS and 3-glycidoxypropyltrimethoxysilane, and deposited onto an Aluminum 5083 alloy. The multifactor analysis of variance (ANOVA) analysis method was employed [21], and pull-off tests assessing the adhesion strength of coatings to the metal substrate were utilized as a response.

In further research [22], the same group exploited DoE based on Taguchi orthogonal design and the ANOVA analysis method to optimize the same sol-gel-derived hybrid organic-inorganic coatings deposited onto an Aluminum 5083 alloy, using the corrosion current density as a response.

Yahyaei et al. [23] investigated the impact of diverse parameters (namely, hydrolysis ratio, molar ratio of precursors, weight percentage of inorganic to organic part, post-curing temperature, and time) on sol-gel derived hybrid UV-curable coatings synthesized from UV-curable urethane acrylates (organic phase) and TEOS and 3-methacryloxypropyl trimethoxy silane (inorganic phase). For this purpose, again, a DoE based on Taguchi orthogonal design was used to identify the best experimental conditions for obtaining hybrid systems with high abrasion resistance and, at the same time, high transparency.

Despite the potential and reliability of DoE, this methodology has rarely been utilized in hybrid organic-inorganic systems. Indeed, sol-gel systems involve many operative conditions (e.g., temperature, pH, chemistry of the sol-gel precursors, sol-gel routes, sol-gel recipes, …), which limit an effective DoE exploitation. As a consequence, it is possible to use DoE for identifying the necessary experimental tests, though their number is still quite high and tests are still time-consuming. To try to fix this issue, during the last five years, a new approach based on the use of AI [24–26] through the exploitation of ML methods [27,28] has emerged (and is emerging). Although it is still in its infancy (Figure 1), this new methodology seems very promising and intriguing, specifically referring to materials science and technology. This new approach involves teaching AI how to exploit already existing experimental data (even incomplete) related to material systems similar to the one being explored to predict key parameters of this latter, minimizing the error while maximizing their reliability.

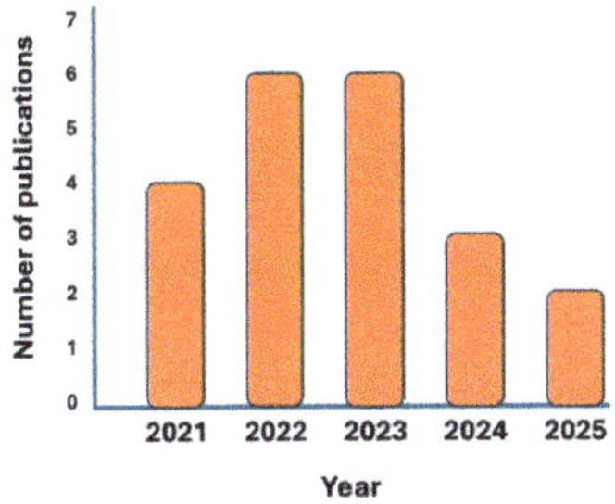

Figure 1. Number of publications per year dealing with "sol-gel" AND "machine learning" (AND is the Boolean operator; source: Web of Science, updated to 8 January 2025).

This review work aims to summarize the research efforts spent so far on the use of AI/ML methods in the field of sol-gel-derived hybrid materials, underlining the current achievements and providing the reader with some perspectives about the development and implementation of this new research approach. To this aim, first, the fundamentals of AI and ML tools will be described; then, the main research outcomes on the use of AI and ML tools for predicting some key properties of hybrid organic-inorganic sol-gel-derived systems will be discussed.

2. Artificial Intelligence and Machine Learning in Materials Science: An Overview

The next paragraphs will summarize the basic concepts of AI and ML tools for those who are not familiar with them, demonstrating that, from an overall point of view, this new strategy can be successfully applied to materials science.

ML is changing the world and science, opening new avenues for the development of innovative and functional materials. Starting from atomistic behavior and physicochemical data, it is possible to predict selected material properties using specific models [29]. Such models can be complex algorithms employed in imaging or material characterization, which can provide data-driven insights in the analysis of material properties and shed light on trends and not-obvious correlations [30]. Predictive models are fostering material discovery and optimization, also enabling the preparation of new compounds with customized features in diverse fields [31]. While, on one hand, the application of wide datasets and ML is making it possible for computers to learn as they progress, on the other hand, it is removing the need for long procedures based on trial and error, expensive routes, and dangerous synthesis steps along the fabrication of products. ML algorithms can recognize patterns, help in risk assessment, adapt over time, and extrapolate useful insights for the design of experiments [32]. ML is a subfield of computer science dealing with AI, and it mainly searches to reveal patterns from data to enhance performance in different activities.

2.1. Supervised Learning Approaches

During the design of a new material, many limitations or constraints can be found; thus, ML is often employed to overcome these adverse factors. In this context, supervised learning is identified as the most representative form of ML [33]. This learning approach involves two steps: the algorithm reveals patterns in the training dataset, which may consist of labels or sample sets, and then it converts patterns into a mathematical function, namely the model, during the training stage. Supervised learning models are generally implemented for predictions about samples that the models have not encountered in the inference stage [34]. To give an example, a typical learning task, following a supervised approach, is the classification problem, in which the learner approximates the behavior related to a function [35]. Training and test stages represent the crucial points of an ML process. In the training phase, the algorithm is built by the exploitation of examples gathered from the training dataset as input, so the acquisition of features from the learner or learning model can take place. Once labeled data are available for training, supervised learning can be applied, for example, in regression (e.g., predicting the prices of products on the market) and classification (e.g., image classification) tasks [36]. The training of linear regression, decision trees (DTs), support vector machines (SVMs), logistic regression, random forests (RFs), and neural network (supervised ones) algorithms is performed on labeled data, thus learning the mapping between the input parameters and their related outputs (see Figure 2) [33].

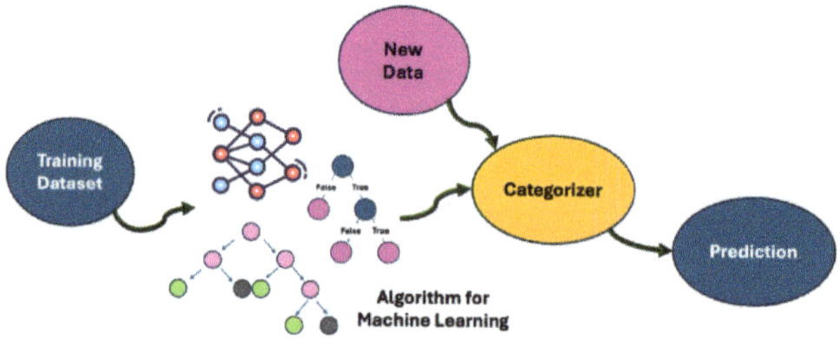

Figure 2. Supervision approaches for the learning process.

In artificial neural networks (ANNs), a hidden layer is a layer composed of artificial neurons, which is different from both the input layer and the output layer, as it appears clear in multilayer perceptrons. Hidden layers transform inputs into a prediction or outputs through weighted connections and activation functions. Hyperparameters, which control the learning process of an ML model, are set before the model starts its training and are different from the internal values learned during the training. The optimization of the hyperparameters allows for the maximization of the model's performance. In a typical ML process, the biases and weights regulate the input of data flow (forward propagation) through the layers of a neural network. Biases and weights are started and then iteratively updated during the training stage by backpropagation [37]. During the backpropagation, the prediction is compared to the actual target value to obtain an error, which is then minimized by the network by adjusting weights and biases. This way, the model learns and improves its prediction accuracy.

Support vector machine (SVM) is a learning model that can be adopted to address and solve regression and classification problems by supervised learning. This model is especially applied to minimize structural risk and find the best hyperplane dividing a dataset into different classes. Its parameters (e.g., kernel and penalty) influence the effectiveness of prediction models. SVMs can employ a kernel trick to convert data into higher dimensions to optimize the splitting of data when these latter are not linearly separable [38]. For this reason, SVMs are more effective and accurate when: (i) the classes are very well separated, (ii) they are implemented for classification tasks in high-dimensional spaces (e.g., text categorization, bioinformatics), and (iii) they are employed to estimate implicit functions [39,40].

2.2. Unsupervised Learning Methods

In materials science, the possibility to look through extensive quantities of data by using ML tools is becoming increasingly important. Unsupervised learning searches for features without manual analysis or explicit supervision. Such a type of learning is useful to find structure or hidden patterns within unlabeled data [41]. To handle customer segmentation or data compression, algorithms based on unsupervised learning (e.g., K-means, principal component analysis (PCA), and generative adversarial networks) are valuable tools [42].

2.3. Deep Learning Methodologies and Bayesian Methods for Optimization

Deep learning is composed of interconnected classifiers based on activation functions and linear regression [43]. Among the existing ML approaches, shallow learning and deep learning differ in their ability to process complex data. More specifically, shallow learning applies models made of one or two hidden layers (i.e., nonlinear feature conversion layers) to handle tasks with simple relationships in data. On the other side, deep learning employs more complex models with multiple hidden layers to learn intricate patterns in data. The most used shallow models are Gaussian mixture models, support vector machines, and logistic regression [44]. For pattern recognition tasks (e.g., image recognition, speech recognition), deep learning is the most appropriate, as it allows for easy management of huge amounts of data. The most frequently employed deep learning algorithms are recurrent neural networks for sequential data, conventional neural networks for image-related tasks, and generative pre-trained transformer models for natural language processing [45]. Bayesian optimization methodology allows for setting parameters for the optimization of the decision-making process by implementing an objective function. Bayesian methods (e.g., Bayesian linear regression and Bayesian networks) are mostly used to model uncertainty based on evidence and objective functions by employing probability [46]; therefore, they are suitable for tasks dealing with decision-making under probabilistic reasoning or in cases where assumptions necessitate being included into the model, for example, in natural language modeling [47,48]. Bayesian optimization strategy optimizes black-box functions (i.e., functions where the internal code and inner workings are hidden; this means that the only interaction of the user with the function concerns the feeding of input and the observation of the outputs), and it represents a valuable tool to find optimal configurations for ML models, especially when the analytical form of the functions is unknown, derivatives are not reliable, and evaluations are expensive. As an example, Gaussian process regression (GPR) is a Bayesian and non-parametric approach to regression, which is mostly appropriate to predict an estimation of the uncertainty, obtain a flexible data-driven model, and model functions with unavailable derivatives.

2.4. Artificial Intelligence in Materials Science and Property Prediction

ML is mostly employed in materials science to predict a material's properties or design new routes of preparation. The algorithm for the prediction is generally built based on an extensive amount of material data. The input dataset is used to train the algorithm, which can then forecast the features of new and untested materials (or intermediates), also along with their synthesis. The proper algorithm can correlate patterns behind the data collected during the experimental work or available in the literature. The possibility to define the future properties of novel materials during their design or manufacturing allows for the identification of the most profitable routes of preparation, screening steps with fewer endeavors, and cost-saving decisions. Thus, the task of discovering the optimal compositions to obtain materials with desired properties can be faced by the ML approach. ML algorithms can be optimal to search for non-linear connections involving several variables concerning the synthesis and characterization of polymers, 2D materials, organic-inorganic structures, and multi-component compounds [49,50]. A huge number of properties have been predicted by ML using wide input datasets (e.g., made of different physico-chemical properties), such as defect energetics [51], glass transition temperatures [52], mechanical properties [53], thermal conductivity [54], and catalytic action [55]. Figure 3 displays how AI finds the way to be applied in material synthesis.

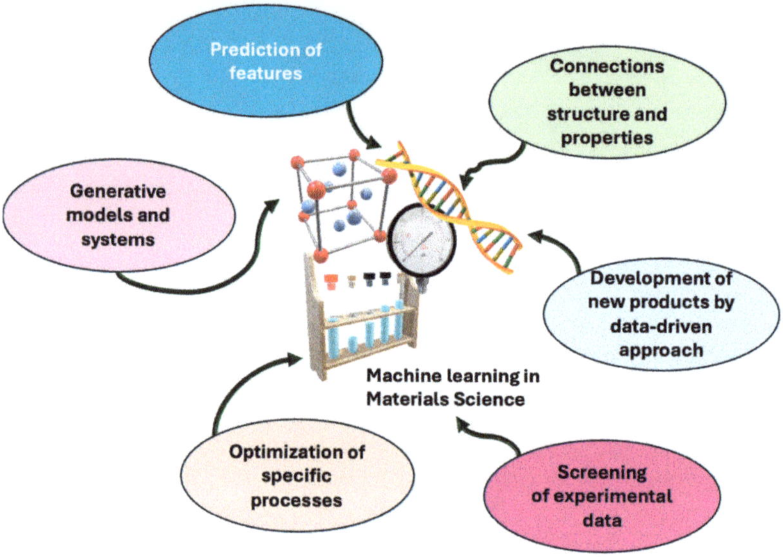

Figure 3. Main routes for exploiting artificial intelligence in materials science.

2.5. Other Machine Learning Algorithms and Statistical Tools for Data Optimization

Adaptive boosting is a well-known ML method that is commonly employed for classification tasks, but it can also be useful for regression [56]. This method forms strong classifiers through the combination of multiple weak learners. Boosted regression trees, or gradient boost (GB) regression trees, are used to connect decision trees and promote the performance of regressions and classifications [57]. Similarly to adaptive boosting, GB regression trees are built by weak learners (e.g., small regression trees) to forecast a continuous target variable: practically, each new tree tries to correct the errors of the combined existing trees. K-nearest neighbors (K-NN) is a non-parametric algorithm applied for both regression and classification, which makes predictions based on the closeness of data points [58]. K-NN starts from the assumption that similar data points are located close to each other in feature space; therefore, such an algorithm stores the whole training dataset and performs the predictions considering the majority class or the average value of the K closest training examples to the input point. Long short-term memory (LSTM) algorithms are used to counter the issue related to the vanishing gradients in deep learning [59]. They are recurrent neural network architectures able to handle prediction tasks like time series forecasting and natural language processing. LSTM finds a wide application with sequential data and in tasks where previous time steps affect future predictions. Finally, among dimensionality reduction (i.e., reduction in input features in datasets, preserving as much variation as possible) techniques, the principal component analysis is frequently implemented in data science to transform large datasets into a set of orthogonal axes (i.e., principal components), which extract the most important variance within the data [60].

This results in a better visualization of high-dimensional data and in an improvement of ML models' performance.

2.6. Statistical Metrics and Performance Parameters for Prediction Evaluation and Data Correlation

Several statistical metrics are generally employed to quantify the performance efficiency of prediction models [61,62]. These metrics give a clear idea of the accuracy and errors of the models in different tasks (e.g., classification, regression). The application of regression models allows for predicting continuous values. Mean absolute error (*MAE*) can be used to measure how close the actual values are to the predicted ones, as it provides the absolute differences between both values:

$$MAE = \frac{1}{n}\sum_{i=1}^{n}|y_i - \hat{y}_i| \quad (1)$$

where y_i is the actual value, $\hat{y}_i$ is the predicted value, and n is the number of data points. Good prediction accuracy corresponds to small *MAE* values. Mean squared error (*MSE*) can be a valuable alternative to *MAE*:

$$MSE = \frac{1}{n}\sum_{i=1}^{n}(y_i - \hat{y}_i)^2 \quad (2)$$

However, *MSE* emphasizes larger errors more than smaller ones because of the squaring, and it is more sensitive than *MAE* to outliers. Root mean squared error (*RMSE*) represents the square root of *MSE*:

$$RMSE = \sqrt{\frac{1}{n}\sum_{i=1}^{n}(y_i - \hat{y}_i)^2} \quad (3)$$

This statistical metric transforms the units to match those of the original data. Also, lower values of *RMSE* indicate better accuracy. *R*-Squared (R^2) indicates the proportion of the variance included in the dependent variable that is possible to predict from the independent variable:

$$R^2 = 1 - \frac{\Sigma(y_i - \hat{y}_i)^2}{\Sigma(y_i - \overline{y})^2} \quad (4)$$

where $\overline{y}$ is the mean of actual values. R^2 can assume values ranging from 0 to 1, with 1 referring to a very reliable prediction. Particularly, a high R^2 reveals that the model well explains most of the variance in the target. Mean absolute percentage error (*MAPE*) can be very useful; it immediately gives an idea about the prediction accuracy of the model as a percentage:

$$MAPE = \frac{100}{n}\sum_{i=1}^{n}\left|\frac{y_i - \hat{y}_i}{y_i}\right| \quad (5)$$

The adoption of such a statistical metric is suggested when the actual values are very small. Due to their definitions, *MSE*, *MAE*, and R^2 are mostly suitable for regression tasks. Concerning classification tasks, the final objective is to predict discrete labels, and thus the accuracy factor is generally employed to evaluate the prediction performance of the algorithm:

$$\text{Accuracy} = \frac{\text{True Positives} + \text{True Negatives}}{\text{Total Instances}} \quad (6)$$

High accuracy means that most of the predictions performed by a specific model are correct [63]. Kling-Gupta efficiency (KGE) is a metric applied to study the performance of continuous time series models through the comparison of simulated values to observed ones [64]. KGE enables the contributions of different components of model performance to be separated, allowing the reasons for good or poor performance to be understood. *R*-value, also called the correlation coefficient, measures the strength of the linear relationship linking two variables. This statistical metric is used in both correlation analysis and regression to estimate the degree of relation between two variables. The strength of the linear relationship between the variables increases with the *R*-value's proximity to either +1 or −1 [65].

3. Current State-of-the-Art in the Use of AI/ML Approaches for Predicting Key Properties of Hybrid Sol-Gel-Derived Materials

The use of ML in the development of new materials is expanding. Sol-gel chemistry enables the preparation of materials, especially glasses and ceramics, by transforming a "sol" into a "gel" through hydrolysis and condensation reactions. This method allows for the synthesis of various systems, including nanoparticles, thin films, and monolithic objects, often at room temperature. Such types of materials are among the most commonly explored and fabricated sol-gel products. In addition to the prediction of materials' properties and advantages under operative conditions, ML models can contribute to emphasizing the valuable characteristics of the sol-gel method, namely: mild synthesis conditions, versatile chemistry, and high sustainability.

3.1. Sol-Gel-Derived Aerogels: Machine Learning Prediction of Their Physico-Chemical Features

Silica aerogels are porous sol-gel systems exhibiting unique characteristics, such as ultra-low density, high surface area, low thermal conductivity, and high porosity. To describe silica aerogels at the nanoscale and their related fractal structure, fractal geometry is generally employed. Abdusalamov et al. proposed an ANN to predict the fractal dimension of silica aerogels and reconstruct their microstructures from a given fractal dimension [66]. The diffusion-limited cluster–cluster aggregation (DLCA) approach was applied to obtain the microstructures [67]. In the forward propagation model, DLCA network structures were used as input parameters to train the ANN. The latter was able to predict the fractal dimension for any given set of input parameters with an accuracy of $R^2 = 0.973$. On the other side, in back-propagation mode and by including model constraints in the inversion algorithm, the same ANN was inverted and employed to predict input DLCA model parameters for a desired fractal dimension with an error of 2% [66].

ANNs are not the only ML tools able to support the development and characterization of such materials. As some of the major applications of silica aerogels are as electrodes in the field of energy conversion and storage and high-end adsorbers, Scherdel et al. proposed the implementation of linear regression and GPR ML models to predict the solid-phase thermal conductivity of silica aerogels from a structural dataset [68]. The research group built the structural input dataset from small-angle X-ray scattering data and physical properties of the aerogels, also using different data subsets as predictors, based on different states of synthesis (i.e., wet and dry), to evaluate the model performance [68]. To give an idea of model evaluation, after supercritical drying, the linear regression model gave an RMSE of ~2 for the training and ~10 for the prediction of the solid-phase thermal conductivity. Considering that the morphology of gels changes a lot during the processing (e.g., supercritical drying), GPR models combined with SAXS data led to the best prediction.

As the properties of silica aerogels are strongly related to the synthesis pathway, it would be beneficial to predict the final features of such materials based on experimental data and operative conditions. In this context, Walker et al. developed a silica aerogel graph database (regarding the synthesis conditions of 10^3 silica aerogels) and a supervised ML neural network regression model to better understand and visually display the relationship between synthesis conditions and final properties of aerogels [69]. By following a validation test, the model maps could predict the aerogel surface area from processing and optimal synthesis conditions with an error lower than 5%, while the graph database was able to reduce time and experimental dimensionality during the preparation of high surface area silica aerogels for application in thermal insulation or sorption media. The silica aerogel graph database was used as an input dataset after removing statistical outliers and aerogels with high prediction errors [69]. This data cleaning process could enhance the model, leading to the prediction of aerogel surface area with an average error of 109 ± 84 m^2/g ($112\% \pm 72.6\%$). The normalized predicted versus actual graph for the surface area values of aerogels in the model, using K-fold cross-validation (i.e., a resampling and sample splitting method that uses different portions of the input data to test and train the model on different iterations), gave R^2 of 0.731, *MSE* value of 0.014, and *RMSE* of 0.118 [69].

Despite the excellent properties of silica aerogels, their mechanical brittleness results in significant challenges. It is well known that surface area, pore volume, density, and thermal conductivity strongly influence the final properties of silica aerogels [70]. However, the silica content may also remarkably affect the performance of such aerogels. Chaouk et al. explored the feasibility of PCA to investigate the influence of silica content on the physicochemical properties of aerogels [71]. Also, cross-correlation analysis was employed to find the relationships between surface area, pore volume, density, and thermal conductivity. PCA results confirmed that silica content is the prominent parameter impacting the aerogel properties, in agreement with the first principal component giving a positive correlation (R^2 = 94%) with the silica content. By avoiding long screening experimental procedures, the research group could demonstrate that higher silica contents correspond to lower thermal conductivity, surface area, and porosity [71]. Finally, the outcomes of the second principal component suggested the thermal conductivity as crucial in influencing the final properties of aerogels, especially in samples with moderate/high silica loadings. Chaouk et al. applied PCA to reduce the dimensionality of the dataset and effectively understand the complex interaction between surface area, porosity, density, thermal conductivity, and silica content of aerogels. PCA results were found optimal, as the statistical method could capture 97.45% of the total variance [71].

3.2. Sol-Gel Derived Hybrid Nanofluids: Machine Learning Prediction of Their Physico-Chemical Properties

Conventional heat transfer liquids (e.g., water, ethylene glycol) show lower thermal conductivity than solid particles. To enhance the thermophysical properties of traditional fluids, the scientific community has proposed different solutions, for example, turbulators and nanofluids. Nanofluids can be prepared by suspending metallic, ceramic, and non-metallic nanoparticles in a base fluid and can guarantee high heat transfer performance [72,73]. Researchers are progressively looking for new, effective, stable, and well-designed nanofluids suitable for fluids with specific density and viscosity. In this context, Said et al. produced, starting from graphene oxide (GO), rGO-Fe_3O_4-TiO_2 ternary hybrid nanocomposites by sol-gel methodology to obtain ethylene glycol-based stable nanofluids, whose viscosity and density were deeply investigated through the variation in temperature and volume concentration [74]. The use of analytical methods for the optimization of viscosity and density of the rGO-Fe_3O_4-TiO_2 hybrid nanofluids may involve long operations to handle a large volume of nonlinear experimental data. To avoid such a time-consuming and expensive procedure, Said et al. employed advanced ML techniques like boosted regression tree (BRT), SVM, and ANNs to predict the influence of temperature and volume concentration on the viscosity and density of rGO-Fe_3O_4-TiO_2 hybrid nanofluids. To have an idea about the prediction performances, statistical errors were calculated to compare predicted values with experimental outcomes [74]. The *R*-values of BRT-based density (0.9989) and viscosity (0.9979) were higher than those of the ANN-based and SVM-based prediction models. These results, together with the values of R^2 (BRT > ANN > SVM) and KGE (BRT > ANN > SVM), demonstrated the higher efficiency of BRT-based prognostic models. Also, the BRT-based models reported lower *MAPE* and *RMSE* than both ANN and SV algorithms. Finally, the research group highlighted that BRT, ANN, and SVM could accurately predict the experimental values of density and viscosity of ternary hybrid nanofluids along wide ranges of temperature and nanoparticle concentration ratio [74].

Among its possible uses, GO can be adopted for the fabrication of water base hybrid nanofluids by combining it with suitable metal oxides. Lately, Kanti et al. prepared hybrid nanofluids by mixing CuO and GO at different ratios. CuO and GO were synthesized by sol-gel chemistry, and the effects of pH on thermal conductivity, stability, and viscosity of hybrid nanofluids were studied for each formulation and composition in a specific range of temperature [75]. The experimental results revealed that these hybrid nanofluids may be promising in solar power facilities, where ethylene glycol or molten salts often require the addition of proper nanoparticles to improve their thermophysical features. Bayesian optimized support vector machine (BoA-SVM) and Bayesian optimized boosted regression trees (BoA-BRT) were applied as ML models for the prediction of thermal conductivity and viscosity of hybrid nanofluids. The prognostic capabilities of models were evaluated by several statistical indices (e.g., *MSE*) [75]. The correlation in terms of R^2 between observed and predicted values of thermal conductivity and viscosity was in the range of 0.9923–0.9922, while the prediction error, measured by MSE, was quite low (between 0.0000085 and 0.00354). BoA-BRT and BoA-SVM were proved as robust forecasting techniques for the prediction, with BoA-SVM (R^2 = 0.9992) showing slightly superior efficiency than BoA-BRT (R^2 = 0.9976). The research group demonstrated that ML can be useful to give a forecast about potential thermophysical parameters of hybrid nanofluids under a wide range of operative conditions, also reducing the implementation of expensive technologies, which are usually employed to test heat transfer fluids [75].

As additional proof that the combined use of sol-gel chemistry and ML can be the needle of the scale in the development of new hybrid nanofluids, Kanti et al. proposed water-based hybrid nanofluids composed of Al_2O_3 and GO at different mixing ratios, also studying their thermophysical characteristics and dispersion stability. Al_2O_3 and GO nanoparticles were synthesized by sol-gel routes and Hummer's method [76]. The stability of the hybrid nanofluids was tested by the incorporation of three different surfactants, while the effects on both thermal conductivity and viscosity were measured at different temperatures (30–60 °C) and filler concentrations (0.1 to 1 vol.%) [76]. The research group found that mainly the inclusion of GO content could increase the thermal conductivity and the viscosity of hybrid nanofluids. Finally, regression models (i.e., a Bayesian optimized support vector machine (BoA-SVM) and a Bayesian optimized wide neural network (BoWNN)) were developed to predict the thermal conductivity and the viscosity of hybrid nanofluids. The prediction of the thermophysical properties achieved a good prognostic efficiency of 97.15–99.91% [76].

3.3. Machine Learning Assisted Sol-Gel Methodologies for the Preparation and Characterization of Functional Polymeric Products

Among the possible materials made by sol-gel chemistry, functional nanostructures composed of different inorganic phases can be employed to fabricate low-energy surface coatings [77]. These coatings are mostly used to generate hydrophobic and superhydrophobic surfaces on a single substrate. In this context, ML tools can predict the optimal set of parameters to obtain hydrophobic surfaces for a given substrate. Manoharan et al. developed an ANN to predict the different factors affecting the anti-wetting behavior and surface contact angle of coated printable paper substrates [78]. The research group used the outcomes of this prediction to optimize the coating method (i.e., dip coating, spray coating, spin coating, and inkjet printing), the number of deposited layers, the selection of proper silane (i.e., trimethoxymethylsilane, TEOS), and the sol-gel-derived material applied during the deposition stage. Manoharan et al. exploited base material, initial contact angle, and final contact angle on the front side and back side as input data for the ANN, which was trained for up to 5 cycles until the convergence ($R \approx 1$) of all parameters. The ANN implemented a tangent sigmoid as a transfer function and the Levenberg–Marquardt algorithm (i.e., an algorithm used to solve non-linear least squares problems), involving Gauss Newton and Gradient methods, as the optimization technique. The contact angle resulting from the predicted parameters was found to be similar to the actual values, with a maximum error of ±4.5° [78].

To be suitable for several industrial applications, polymeric composites must pass specific requirements in terms of fire behavior. As most polymers are highly flammable, the exploitation of flame retardants is generally needed [79]. As an example, the incorporation of magnesium hydroxide nanoparticles into the polymer matrix allows for a reduction in the heat release rate and smoke emission of final composites, also lowering their flammability. To test the fire performance of polymeric materials, a long series of disruptive and time-consuming fire tests must be performed to find the optimal formulation for a specific application [80]. ML tools can help to expedite the development of new flame-retardant polymeric composites or support their characterization by predicting key fire parameters. Recently, Bifulco et al. synthesized an in-situ hybrid $Mg(OH)_2$-epoxy nanocomposite through sol-gel methodology [81], also predicting its heat release capacity through the implementation of a made-on-purpose fully connected feed-forward ANN (see Figure 4) [82]. From the experimental side, nanocomposites incorporating ~5 wt.% of in-situ generated $Mg(OH)_2$ nanoparticles exhibited a significant decrease in the HRC (around −34%), measured by performing a certain number of pyrolysis combustion flow calorimetry tests, and in some cone calorimetry parameters (e.g., peak of heat release rate (pkHRR), heat release rate, and total smoke release) [82]. The developed ANN-based system could predict the HRC of the prepared $Mg(OH)_2$-epoxy nanocomposite, giving very low error values (i.e., MAE and RMSE equal to 145.6 and 186.1, respectively).

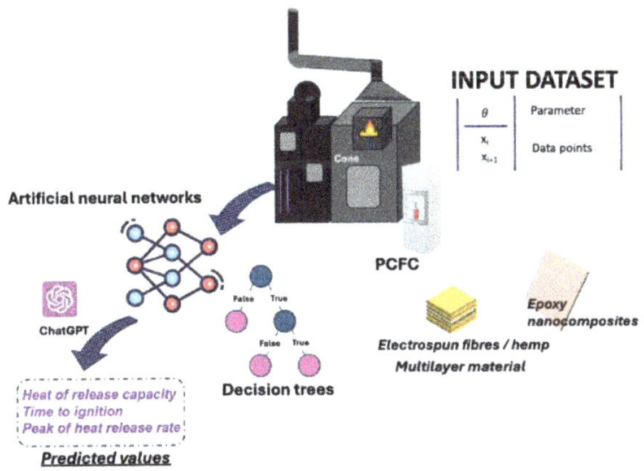

Figure 4. Development of hybrid epoxy nanocomposites and multifunctional textiles and prediction of their fire properties.

Sol-gel chemistry has also been combined with a DCNN (deep convolutional neural network) algorithm: this latter applies a convolution operation over one-dimensional sequence data, which are commonly adopted to analyze temporal signals or text. A one-dimensional convolutional neural network (1DCNN) in deep learning is specifically designed for processing one-dimensional sequence data [83]. To this aim, Ma et al. developed a system for the shelf-life estimation of food products, namely the freshness-related information of salmon for consumers [84]. The system was fabricated by incorporating a dye-containing, optically changeable colorimetric sensor into a sol-gel functionalized paper sensor using a DCNN model based on images for the freshness estimation. The cross-reactive paper sensor array was obtained by employing different dyes and treating paper with sol-gel particles (synthesized from (3-aminopropyl) triethoxysilane -APTMS- and TEOS) and indicators (e.g., ammonia gas indicator, pH indicator, indicators for volatile organic compounds, and oxidation indicator) [84]. The augmentation by ML allowed for setting the limit for ammonia detection at ~17.1 ppm and the accuracy up to 99.2% for the non-destructive estimation of the salmon freshness (i.e., visual and olfactory), in agreement with the requirements of food supply chains and scalable batch production [84].

Another interesting application of AI/ML tools refers to electrospinning and sol-gel-derived nanostructures, which can be used to develop polymeric nanofibers with several functional properties. In this context, Bifulco et al. prepared a multilayer material made of coated hemp blankets with cross-linked electrospun polyvinylpyrrolidone-silica blankets and TiO_2 nanoparticles (Figure 4) [85]. This MM showed good tensile properties and could work not only as a smoke suppressant but also as a fireproof fabric. The assessment of the fire properties is usually time-consuming and requires the preparation of a huge number of precious samples. Despite the incomplete input dataset consisting of fire parameters related to sol-gel-treated textiles, and thanks to the support of ChatGPT (GPT-4o version), the research group implemented ANNs and decision trees to predict the time to ignition (TTI) and the pkHRR of the MM. The accuracy of the prediction was evaluated by the calculation of MAPE values, considering the experimental data and the predicted values. Decision trees gave *MAPE* values of 0.431 (~43%) and 0.336 (~34%) for TTI and pkHRR, respectively. Conversely, *MAPE* values assessed by the outputs of ANNs were 0.468 (~47%) for the TTI parameter and 0.404 (~40%) for the pkHRR parameter [85]. Based on these results, for both algorithms, a reasonable predictive capability was found [86]. However, decision trees provided the lowest *MAPE* values, hence becoming preferable and more reliable in case of some missing input data.

Hybrid sol-gel materials combine the best features of the organic and inorganic phases in a single material due to the peculiarities of the resulting co-continuous network. In this respect, Lin et al. demonstrated that sol-gel chemistry and ML can be applied to obtain large hybrid monoliths, made of silica/poly (tetrahydrofuran), with good mechanical properties as well as resistance toward cracking phenomena [87]. These latter are commonly triggered by the combined action of capillary and shrinkage stresses inside the monoliths, which mostly take place during the drying procedure. The shrinkage along different drying processes was assessed by collecting computational scribe calculations on X-ray micro-computed tomography images. A filtered back-projection algorithm was applied to reconstruct captured projections into a 3D image matrix. A random forest voxel classifier (i.e., an ML algorithm) was implemented to segment hybrid samples within the 3D images [87,88].

3.4. Sol-Gel Derived Metal Oxides, Organic Structures, and Inorganic Materials for Several ML-Optimized Technologies

Many sol-gel techniques can be employed for the manufacturing of coatings able to confer very low wettability and chemical resistance to different substrates [89]. To give an example, in the oil and gas sector, the wettability of surfaces in pipelines affects the corrosion rates; therefore, several strategies are usually adopted to limit the occurrence of material deterioration phenomena [90]. Akbarzadeh et al. obtained sol-gel coatings starting from TEOS, methyltriethoxysilane, and glycidoxypropyltrimethoxysilane as precursors (Figure 5) [91]. The research group incorporated different amounts of oxidized multi-walled carbon nanotubes (O-MWCNTs) to modify the corrosion resistance properties and hydrophobicity, and increase the barrier effect of the silane-based nanocomposite coatings, also checking the performance through electrochemical impedance spectroscopy and wettability measurements. By using a single hidden-layer perceptron and the Levenberg–Marquardt algorithm optimization procedure as a learning algorithm, an ANN model was constructed to predict the imaginary part of the impedance [91]. The immersion time of mild steel panels in the sol-gel silane solution by dip-coating, the different loadings of O-MWCNT in the solution, and the real part of the impedance were implemented as input data. The accuracy of the optimized model was calculated by comparing the experimental data with the predicted parameters. High O-MWCNT loadings in the silane-based coatings resulted in a decrease in the corrosion current density, as confirmed by polarization curves and AC impedance spectra. The ANN model demonstrated highly reliable predictive efficiency, as the statistical errors, including *MSE* and *MAE*, were 0.00099606 and 0.0106, respectively. Also, the R^2 value was found to be around 0.9879, revealing the absence of scattering in the entire data prediction range [91].

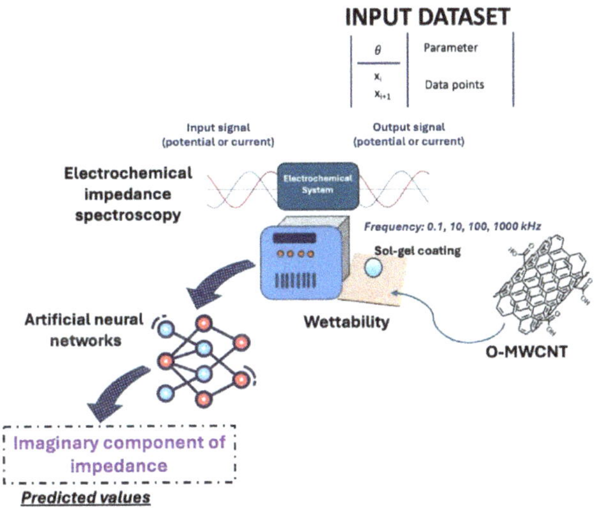

Figure 5. Development of sol-gel coatings containing O-MWCNTs and prediction of the imaginary part of the impedance for corrosion resistance technologies.

Further, AI/ML tools can support sol-gel chemistry along with the synthesis of specific nanostructures that are crucial for the preparation of biosensors [92]. Biosensors' receptors include enzymes, nucleic acids, and microorganisms, among others. To open new perspectives in the fabrication of nanomaterial-based sensors, Zalke et al. proposed an ML-assisted screen-printed and non-enzymatic biosensor for the quantification of urea concentrations, which is very important during medical intervention [93]. For the detection of urea, the biosensor was tailored by an MWCNT-ZnO nanocomposite functionalized with CuO micro-flowers (MFs). CuO-MFs were prepared by the sol-gel approach starting from copper nitrate trihydrate as a precursor. To have an idea about the sensor's performance for the detection of urea, cyclic voltammetry at different scan rates was applied to assess the dependence of peak currents on analyte concentration. The non-enzymatic biosensor exhibited good sensitivity and a satisfactory linear range of operation [93]. Regression-based ML models were employed to predict CV parameters for the optimization of the measured experimental concentrations of urea. To improve the sensor's accuracy and determination of the urea concentration, linear regression, DT, K-nearest neighbor (KNN), RF, adaptive boosting (AdBoost), and GB models were used, and their outcomes were compared by different regression metrics. ML models outperformed linear regression in predictive accuracy metrics, e.g., *RMSE* and R^2. For example, the KNN model gave a significantly lower *RMSE* (0.0004) than linear regression (0.0006), together with a higher (0.981) R^2 compared to linear regression (0.953) [93]. Thus, the research group demonstrated that the KNN model could well predict optimal CV parameters and thus the sensor's response to different urea concentrations, not only providing a better fit of the data but also capturing more variance.

Water is a precious resource for humans, and its proper management is getting increasingly complicated, as new pollutants and persistent contaminants are constantly found during analysis [94]. In the field of water decontamination, green strategies or waste-to-wealth approaches allow for the synthesis of functional nanoparticles by the sol-gel method [95]. ZnO nanoparticles can be very effective in addressing groundwater contamination and countering microbial cells. Chelghoum et al. obtained ZnO-based nanoparticles by using a sol-gel route and lemon peel aqueous extract [96]. ZnO-based nanoparticles were employed to photodegrade quinoline yellow dye by sunlight irradiation while studying their antioxidant and antibacterial properties. The nanoparticles exhibited a notable photoactivity, also revealing high efficiency as antioxidants and against Gram-positive bacterial strains. The photodegradation action was mainly affected by the initial concentration of the quinoline yellow, the pH, and the catalyst dosage. To predict the photodegradation rates of quinoline yellow, the research group developed a GPR combined with an optimization Lévy flight distribution algorithm [96]. The implementation of an ARD-Exponential kernel function (i.e., a tool enabling the non-linear transformation of data without explicitly performing such transformations, also mapping input data into a high-dimensional feature space) allowed us to achieve high accuracy. The prediction model gave high correlation coefficients (i.e., close to 1) between predicted values and the actual ones and low *RMSE* values along all phases (i.e., 0.0160, 0.0250, 0.0244, and 0.0179 during training, testing, validation, and for the whole dataset, respectively), confirming good reliability in forecasting photodegradation rates of quinoline yellow by the synthesized nanoparticles. The prediction of dye photodegradation could avoid time-consuming procedures and waste of resources [96].

Furthermore, nanostructured organic-inorganic hybrids made by ML-assisted sol-gel procedures can represent potential solutions and facilitate new advancements for water-splitting studies [97]. In this context, Koyale et al. improved the photoelectrochemical activity of graphitic carbon nitride (g-C_3N_4) nanosheets through the synthesis of their nanocomposites with metal-organic framework-derived porous CeO_2 nanobars along with ZnO nanorods and TiO_2 nanoparticles [98]. Such synthesis could enhance the photoelectrochemical behavior of g-C_3N_4 by lowering the charge recombination rate. CeO_2 nanobars and g-C_3N_4 nanosheets were prepared starting from their precursors, whereas the sol-gel method was applied to obtain ZnO nanorods and TiO_2 nanoparticles. The binder-free brush-coating technique was deployed to manufacture nanocomposites-based photoanodes, which were analyzed through several electrochemical experiments [98]. The binary g-C_3N_4/CeO_2 nanocomposites, including 20 wt.% CeO_2 nanobars (gC20 nanocomposites), showed a significant enhancement in the current density compared to that of bare g-C_3N_4 nanosheets. A further improvement in the current density was detected using ternary gC20/TiO_2 (gCT50) and gC20/ZnO (gCZ50) nanocomposites: this finding was mainly ascribed to a lower charge transfer resistance and an increased donor density. The research group also predicted the stability (i.e., the values of current density over time) of gCZ50 and gCT50 nanocomposite-based photoanodes by the implementation of a recurrent neural network-based LSTM technique [98]. This technique could effectively forecast the stability of photoanodes, providing useful details about their durability. The performance efficiency of the prediction was evaluated by the calculation of *RMSE* and *MSE* metrics: both showed very low values for the training datasets (80% of the experimental electrochemical observations) and the stability prediction results.

Then, ML, with its data-driven approach, has successfully been exploited in the design of materials for batteries. Owing to such interest, Seitz et al. developed an approach using data obtained from small-angle X-ray scattering measurements for the manufacturing of optimized sol–gel Si-based porous materials for battery anodes [99]. The research group employed a feed-forward ANN connected to a pretrained autoencoder to map parameters (e.g., network topologies) related to the material synthesis to the SAXS curve of the resulting product. The trained network could give information about which parameters allow for a desired SAXS curve [99]. It was found that the prediction error of the ANN model could be reduced by widening the input dataset (during the design of the experiment stage) and training each output variable with an independent ANN algorithm, hence following a multi-step method for the material development.

Recently, the 1DCNN algorithm was also applied to prepare a sensor array able to effectively classify mixed gases. Such a sensor may be promising for developing electronic nose systems with low data processing and high sensitivity. Mu et al. fabricated a ZnO/SnO_2-based micro-electromechanical system gas sensor array by inkjet printing sensing materials onto a micro-hotplate [100]. By combining the use of a pattern recognition algorithm and having time series input lengths as starting data, the sensor could identify formaldehyde, toluene, ammonia, and their binary and ternary mixtures. A sol-gel route was employed to prepare the pure ZnO precursor ink. Unlike traditional ML models, the pattern recognition unit based on the 1DCNN algorithm could avoid the manual feature extraction step, prevent any alteration in the original features of the data, and reduce the experimental work [100]. Furthermore, unlike the best-performing traditional algorithm, i.e., the SVM model, which gave an 80% recognition accuracy of the gas mixtures, the 1DCNN could achieve 99.8%.

To make better use of the systems discussed above, Table 1 summarizes their main features and outcomes.

Table 1. Sol-gel system technologies and outcomes related to the prediction of some specific properties and functions.

Sol-Gel System	Machine Learning and Optimization Tools	Predicted Parameters and Characteristics	Statistical Indices for Efficiency	Ref.
Silica aerogels	ANN	• Fractal dimension • Reconstruct micro-structures	R^2 = 0.973	[66]
Silica aerogels	• Linear regression • Gaussian process regression	Solid-phase thermal conductivity	• RMSE = ~2 for the training • RMSE = ~10 for the prediction	[68]
Silica aerogels	• Supervised ML neural network regression • K-fold cross-validation	Surface area	• R^2 = 0.731 • MSE = 0.014 • RMSE = 0.118	[69]
rGO-Fe_3O_4-TiO_2 hybrid nanofluids	• BRT • SVM • ANN	Influence of temperature and volume concentration on the viscosity and density	R-values of BRT-based density (0.9989) and viscosity (0.9979) higher than those of the ANN-based and SVM-based models	[74]

Table 1. *Cont.*

Sol-Gel System	Machine Learning and Optimization Tools	Predicted Parameters and Characteristics	Statistical Indices for Efficiency	Ref.
Hybrid nanofluids based on CuO and GO at different ratios	• BoASVM • BoA-BRT	• Thermal conductivity • Viscosity	• BoASVM (R^2 = 0.9992) • BoA-BRT (R^2 = 0.9976)	[75]
Hybrid nanofluids composed of alumina (Al_2O_3) and graphene oxide (GO) at different mixing ratios	• BoASVM • BoWNN	• Thermal conductivity • Viscosity	Prognostic efficiency of 97.15–99.91%	[76]
$Mg(OH)_2$-epoxy nanocomposite	ANN	HRC	MAE and RMSE equal to 145.6 and 186.1, respectively	[82]
Sol-gel functionalized paper sensor	• DCNN model based on images • Data augmentation	Non-destructive estimation of the salmon freshness	Accuracy up to 99.2%	[84]
Coated hemp blankets with cross-linked electrospun polyvinylpyrrolidone-silica blankets and TiO_2 nanoparticles	• ANNs • Decision trees • ChatGPT	• TTI • pkHRR	• MAPE values of 0.431 (~43%) and 0.336 (~34%), respectively for the TTI parameter and the pkHRR parameter • MAPE values assessed by the outputs of ANNs were 0.468 (~47%) for the TTI parameter and 0.404 (~40%) for the pkHRR parameter	[85]
Sol-gel coatings containing O-MWCNT nanotubes	• ANN • Levenberg–Marquardt algorithm	Imaginary part of the impedance	• MSE and MAE were 0.00099606 and 0.0106, respectively • R^2 value was found around 0.9879	[91]
Xanthine biosensor based on Co_3O_4 nanoparticles and MWCNTs (tailored by a MWCNT-ZnO nanocomposite functionalized with CuO-MFs)	• Linear regression • DT • KNN • RF • AdBoost • GB	• Optimization of the measured experimental concentrations of urea • CV parameters	KNN model gave lower RMSE (0.0004) than linear regression (0.0006), together with a higher (0.981) R^2 compared to linear regression (0.953)	[93]
ZnO-based nanoparticles	• GPR • Lévy flight distribution	Photodegradation rates of quinoline yellow	RMSE values along all phases (i.e., training (0.0160), testing (0.0250), validation (0.0244), and the whole dataset (0.0179))	[96]
ZnO/SnO_2-based micro-electromechanical system gas sensor array	1DCNN	Classification of mixed gases	The conventional SVM model gave an 80% recognition accuracy of the gas mixtures, while the 1DCNN provided 99.8%	[100]

4. Conclusions and Perspectives

Undoubtedly, combining AI with ML tools offers big potentialities as a valuable, reliable, and efficient strategy for predicting several properties of material systems.

This review paper first set out how AI and ML can be successfully applied to materials science, focusing particularly on hybrid organic-inorganic systems derived from sol-gel processes. Although work in this area is still in its infancy, predicting the properties of these hybrid systems while minimizing errors and maximizing the results' reliability is of extreme interest. This is also relevant in limiting the time and energy spent on experimental investigations. It is worth noting that using a combined AI and ML approach does not mean avoiding experimental testing activities but rather being able to exploit this new strategy to save time and maximize the optimization of research outcomes.

Comparing the potential of AI/ML tools with DoE, it is worth noticing that the two strategies are completely different and rely on a very distinctive approach. DoE is applied to the system under investigation only, trying to minimize the experimental testing necessary to get an acceptable description of the system itself and using a "learning by doing" approach carried out by the investigator. Further, DoE often exploits a linear correlation among the identified variables that are characteristics of the investigated systems.

Conversely, AI/ML tools show the advantage of (i) using also non-linear correlations among the selected variables, (ii) exploiting datasets of material systems (similar to the investigated one) through, e.g., a data augmentation approach or generative models, and (iii) predicting the final properties of a material system by training algorithms on a based input dataset. This is surely a different strategy based on a "learning by doing" approach performed by an AI algorithm properly selected and managed by the operator.

Despite its high potential, this approach still has some limitations that currently prevent its widespread and effective use. From a general point of view, the combined exploitation of AI and ML requires access to a large dataset of properties of similar materials: indeed, the accuracy of predictions about the properties of the material under study depends heavily on the availability of reliable materials data. However, the availability of large datasets is not always guaranteed, which limits the use of this new strategy to a certain extent, even for sol-gel-derived systems. Furthermore, despite the significant research activity that has taken place in recent years, there is currently widespread skepticism surrounding the use of AI in applied sciences, particularly materials science. This is certainly slowing down the development and implementation of AI and ML tools in materials science. Using this new approach also requires cross-cutting skills related not only to sol-gel process chemistry and hybrid organic-inorganic systems but also to neural networks and decision trees. This requires complementary figures to be involved in the research groups.

In conclusion, despite the above limitations, these new approaches will likely see strong further development in the coming years, paving the way for their increasingly "intelligent", reliable, and extensive use.

Author Contributions: Conceptualization, A.B. and G.M.; resources, G.M.; data curation, A.B.; writing—original draft preparation, A.B. and G.M.; writing—review and editing, A.B. and G.M.; supervision, G.M.; funding acquisition, G.M. All authors have read and agreed to the published version of the manuscript.

Funding: This research received no external funding.

Institutional Review Board Statement: Not applicable.

Informed Consent Statement: Not applicable.

Data Availability Statement: No new data were created or analyzed in this study. Data sharing is not applicable to this article.

Conflicts of Interest: The authors declare no conflicts of interest.

Abbreviations

The following abbreviations are used in this manuscript:

Adaptive boosting	AdBoost
Artificial Intelligence	AI

Artificial neural network	ANN
Bayesian optimized boosted regression trees	BoA-BRT
Bayesian optimized support vector machine	BoASVM
Bayesian optimized wide neural network	BoWNN
Boosted regression tree	BRT
Decision tree	DT
Deep convolutional neural network	DCNN
Diffusion-limited cluster–cluster aggregation	DLCA
Gaussian process regression	GPR
Gradient boost	GB
Graphene oxide	GO
Graphitic carbon nitride	g-C_3N_4
Heat release capacity	HRC
K-nearest neighbor	KNN
Long short-term memory	LSTM
Machine learning	ML
Mean absolute error	MAE
Mean absolute percentage error	MAPE
Mean square error	MSE
Multiwall carbon nanotubes	MWCNTs
One-dimensional convolutional neural network	1DCNN
Oxidized multi-walled carbon nanotubes	O-MWCNTs
Peak of heat release rate	pkHRR
Principal component analysis	PCA
Random forest	RF
Root main square error	RMSE
Support vector machine	SVM
Tetraethylorthosilicate	TEOS
Time to ignition	TTI

References

1. Barrino, F. Hybrid Organic–Inorganic Materials Prepared by Sol–Gel and Sol–Gel-Coating Method for Biomedical Use: Study and Synthetic Review of Synthesis and Properties. *Coatings* **2024**, *14*, 425. [CrossRef]
2. Zarkov, A. Sol–Gel Technology Applied to Materials Science: Synthesis, Characterization and Applications. *Materials* **2024**, *17*, 462. [CrossRef] [PubMed]
3. Zhang, Z.; Huang, Y.; Xie, Q.; Liu, G.; Ma, C.; Zhang, G. Functional Polymer–Ceramic Hybrid Coatings: Status, Progress, and Trend. *Prog. Polym. Sci.* **2024**, *154*, 101840. [CrossRef]
4. Letailleur, A.A.; Ribot, F.; Boissière, C.; Teisseire, J.; Barthel, E.; Desmazieres, B.; Chemin, N.; Sanchez, C. Sol–Gel Derived Hybrid Thin Films: The Chemistry behind Processing. *Chem. Mater.* **2011**, *23*, 5082–5089. [CrossRef]
5. Schottner, G. Hybrid Sol− Gel-Derived Polymers: Applications of Multifunctional Materials. *Chem. Mater.* **2001**, *13*, 3422–3435. [CrossRef]
6. Figueira, R.B. Hybrid Sol–Gel Coatings for Corrosion Mitigation: A Critical Review. *Polymers* **2020**, *12*, 689. [CrossRef] [PubMed]
7. Biglari, L.; Naghdi, M.; Poursamar, S.A.; Nilforoushan, M.R.; Bigham, A.; Rafienia, M. A Route toward Fabrication of 3D Printed Bone Scaffolds Based on Poly (Vinyl Alcohol)–Chitosan/Bioactive Glass by Sol–Gel Chemistry. *Int. J. Biol. Macromol.* **2024**, *258*, 128716. [CrossRef] [PubMed]
8. Boukhriss, A.; Roblin, J.-P.; Aaboub, T.; Boyer, D.; Gmouh, S. Luminescent Hybrid Coatings Prepared by a Sol–Gel Process for a Textile-Based PH Sensor. *Mater. Adv.* **2020**, *1*, 918–925. [CrossRef]
9. Balbay, S.; Acıkgoz, C. Anti-Yellowing UV-Curable Hybrid Coatings Prepared by the Sol–Gel Method on Polystyrene. *Prog. Org. Coat.* **2020**, *140*, 105499. [CrossRef]
10. Bifulco, A.; Imparato, C.; Aronne, A.; Malucelli, G. Flame Retarded Polymer Systems Based on the Sol-Gel Approach: Recent Advances and Future Perspectives. *J. Sol-Gel Sci. Technol.* **2022**, *115*, 226–250. [CrossRef]
11. Arya, M.; Heera, S.; Meenu, P.; Deepa, K.G. Organic-Inorganic Hybrid Materials and Architectures in Optoelectronic Devices: Recent Advancements. *ChemPhysMater* **2024**, *3*, 252–272. [CrossRef]

12. Waqar, M.A.; Mubarak, N.; Khan, A.M.; Khan, R.; Khan, I.N.; Riaz, M.; Ahsan, A.; Munir, M. Sol-Gel for Delivery of Molecules, Its Method of Preparation and Recent Applications in Medicine. *Polym. Technol. Mater.* **2024**, *63*, 1564–1581. [CrossRef]
13. Onishi, B.S.D.; Costa, B.H.; Marchiori, L.; de Freitas, B.D.; Pugina, R.S.; Bartlett, J.R.; Carlos, L.D.; Ferreira-Neto, E.P.; Ribeiro, S.J.L. Organic–Inorganic Hybrid Silica Systems: Synthesis Strategies and Optical Applications. *J. Sol-Gel Sci. Technol.* **2025**, *113*, 86–108. [CrossRef]
14. Samuneva, B.; Kabaivanova, L.; Chernev, G.; Djambaski, P.; Kashchieva, E.; Emanuilova, E.; Miranda Salvado, I.M.; Fernandes, M.H.V.; Wu, A. Sol–Gel Synthesis and Structure of Silica Hybrid Materials. *J. Sol-Gel Sci. Technol.* **2008**, *48*, 73–79. [CrossRef]
15. Faustini, M.; Nicole, L.; Ruiz-Hitzky, E.; Sanchez, C. History of Organic–Inorganic Hybrid Materials: Prehistory, Art, Science, and Advanced Applications. *Adv. Funct. Mater.* **2018**, *28*, 1704158. [CrossRef]
16. Gross, S.; Müller, K. Sol–Gel Derived Silica-Based Organic–Inorganic Hybrid Materials as "Composite Precursors" for the Synthesis of Highly Homogeneous Nanostructured Mixed Oxides: An Overview. *J. Sol-Gel Sci. Technol.* **2011**, *60*, 283–298. [CrossRef]
17. Wen, J.; Wilkes, G.L. Organic/Inorganic Hybrid Network Materials by the Sol−Gel Approach. *Chem. Mater.* **1996**, *8*, 1667–1681. [CrossRef]
18. Jiju, A. *Design of Experiments for Engineers and Scientists*, 3rd ed.; Elsevier: Amsterdam, The Netherlands, 2023; pp. i–iii, ISBN 978-0-443-15173-6.
19. Mahmoodian, M.; Arya, A.B.; Pourabbas, B. Synthesis of Organic–Inorganic Hybrid Compounds Based on Bis-GMA and Its Sol–Gel Behavior Analysis Using Taguchi Method. *Dent. Mater.* **2008**, *24*, 514–521. [CrossRef] [PubMed]
20. Najafabadi, A.H.; Mozaffarinia, R.; Rahimi, H.; Razavi, R.S.; Paimozd, E. Sol–Gel Processing of Hybrid Nanocomposite Protective Coatings Using Experimental Design. *Prog. Org. Coat.* **2013**, *76*, 293–301. [CrossRef]
21. Roy, R.K. *Taguchi Method*, 2nd ed.; JUSE Press, Ltd.: Tokyo, Japan, 2010; ISBN 9780872638648.
22. Rahimi, H.; Mozaffarinia, R.; Najafabadi, A.H.; Razavi, R.S.; Paimozd, E. Optimization of Process Factors for the Synthesis of Advanced Chrome-Free Nanocomposite Sol–Gel Coatings for Corrosion Protection of Marine Aluminum Alloy AA5083 by Design of Experiment. *Prog. Org. Coat.* **2013**, *76*, 307–317. [CrossRef]
23. Yahyaei, H.; Mohseni, M.; Bastani, S. Using Taguchi Experimental Design to Reveal the Impact of Parameters Affecting the Abrasion Resistance of Sol–Gel Based UV Curable Nanocomposite Films on Polycarbonate. *J. Sol-Gel Sci. Technol.* **2011**, *59*, 95–105. [CrossRef]
24. Ooi, K.-B.; Tan, G.W.-H.; Al-Emran, M.; Al-Sharafi, M.A.; Capatina, A.; Chakraborty, A.; Dwivedi, Y.K.; Huang, T.-L.; Kar, A.K.; Lee, V.-H. The Potential of Generative Artificial Intelligence across Disciplines: Perspectives and Future Directions. *J. Comput. Inf. Syst.* **2025**, *65*, 76–107. [CrossRef]
25. Gama, F.; Magistretti, S. Artificial Intelligence in Innovation Management: A Review of Innovation Capabilities and a Taxonomy of AI Applications. *J. Prod. Innov. Manag.* **2025**, *42*, 76–111. [CrossRef]
26. Gursoy, D.; Cai, R. Artificial Intelligence: An Overview of Research Trends and Future Directions. *Int. J. Contemp. Hosp. Manag.* **2025**, *37*, 1–17. [CrossRef]
27. Barbierato, E.; Gatti, A. The Challenges of Machine Learning: A Critical Review. *Electronics* **2024**, *13*, 416. [CrossRef]
28. Worden, K.; Tsialiamanis, G.; Cross, E.J.; Rogers, T.J. *Artificial Neural Networks*; Springer: Berlin, Germany, 2023; ISBN 9783031366444.
29. Meier, R.J. A Way towards Reliable Predictive Methods for the Prediction of Physicochemical Properties of Chemicals Using the Group Contribution and Other Methods. *Appl. Sci.* **2019**, *9*, 1700. [CrossRef]
30. Diez-Olivan, A.; Del Ser, J.; Galar, D.; Sierra, B. Data Fusion and Machine Learning for Industrial Prognosis: Trends and Perspectives towards Industry 4.0. *Inf. Fusion* **2019**, *50*, 92–111. [CrossRef]
31. Mishra, R.K.; Mishra, D.; Agarwal, R. An Artificial Intelligence-Powered Approach to Material Design. In *Cutting-Edge Research in Chemical and Material Science*; Bhumi: Kolhapur, India, 2024; Volume I, pp. 62–89.
32. Plathottam, S.J.; Rzonca, A.; Lakhnori, R.; Iloeje, C.O. A Review of Artificial Intelligence Applications in Manufacturing Operations. *J. Adv. Manuf. Process.* **2023**, *5*, e10159. [CrossRef]
33. Osisanwo, F.Y.; Akinsola, J.E.T.; Awodele, O.; Hinmikaiye, J.O.; Olakanmi, O.; Akinjobi, J. Supervised Machine Learning Algorithms: Classification and Comparison. *Int. J. Comput. Trends Technol.* **2017**, *48*, 128–138. [CrossRef]
34. Burkart, N.; Huber, M.F. A Survey on the Explainability of Supervised Machine Learning. *J. Artif. Intell. Res.* **2021**, *70*, 245–317. [CrossRef]
35. Polce, E.M.; Kunze, K.N.; Fu, M.C.; Garrigues, G.E.; Forsythe, B.; Nicholson, G.P.; Cole, B.J.; Verma, N.N. Development of Supervised Machine Learning Algorithms for Prediction of Satisfaction at 2 Years Following Total Shoulder Arthroplasty. *J. shoulder Elb. Surg.* **2021**, *30*, e290–e299. [CrossRef] [PubMed]
36. Shetty, S.H.; Shetty, S.; Singh, C.; Rao, A. Supervised Machine Learning: Algorithms and Applications. In *Fundamentals and Methods of Machine and Deep Learning: Algorithms, Tools, and Applications*; Scrivener Publishing LLC: Beverly, MA, USA, 2022; pp. 1–16.

37. Terlapu, P.; Sadi, R.P.; Pondreti, R.; Tippana, C. Intelligent Identification of Liver Diseases Based on Incremental Hidden Layer Neurons ANN Model. *Int. J. Comput. Digit. Syst.* **2022**, *11*, 1027–1050. [CrossRef] [PubMed]
38. Brereton, R.G.; Lloyd, G.R. Support Vector Machines for Classification and Regression. *Analyst* **2010**, *135*, 230–267. [CrossRef] [PubMed]
39. Ortegon, J.; Ledesma-Alonso, R.; Barbosa, R.; Castillo, J.V.; Atoche, A.C. Material Phase Classification by Means of Support Vector Machines. *Comput. Mater. Sci.* **2018**, *148*, 336–342. [CrossRef]
40. Deng, J. Structural Reliability Analysis for Implicit Performance Function Using Radial Basis Function Network. *Int. J. Solids Struct.* **2006**, *43*, 3255–3291. [CrossRef]
41. Usama, M.; Qadir, J.; Raza, A.; Arif, H.; Yau, K.-L.A.; Elkhatib, Y.; Hussain, A.; Al-Fuqaha, A. Unsupervised Machine Learning for Networking: Techniques, Applications and Research Challenges. *IEEE Access* **2019**, *7*, 65579–65615. [CrossRef]
42. Kalusivalingam, A.K.; Sharma, A.; Patel, N.; Singh, V. Enhancing Customer Segmentation through AI: Leveraging K-Means Clustering and Neural Network Classifiers. *Int. J. AI ML* **2020**, *1*. Available online: https://cognitivecomputingjournal.com/index.php/IJAIML-V1/article/view/49 (accessed on 23 June 2025).
43. Mishra, C.; Gupta, D.L. Deep Machine Learning and Neural Networks: An Overview. *IAES Int. J. Artif. Intell.* **2017**, *6*, 66. [CrossRef]
44. Misra, A.; Vojinovic, Z.; Ramakrishnan, B.; Luijendijk, A.; Ranasinghe, R. Shallow Water Bathymetry Mapping Using Support Vector Machine (SVM) Technique and Multispectral Imagery. *Int. J. Remote Sens.* **2018**, *39*, 4431–4450. [CrossRef]
45. Kumari, P.; Swarnalatha, K.S.; Pandey, P. Deep Learning: Foundations, Architectures, and Transformative Applications. In *Neural Network Advancements in the Age of AI*; IGI Global: Hershey, PA, USA, 2025; pp. 33–98.
46. Shakya, A.; Biswas, M.; Pal, M. Classification of Radar Data Using Bayesian Optimized Two-Dimensional Convolutional Neural Network. In *Radar Remote Sensing*; Elsevier: Amsterdam, The Netherlands, 2022; pp. 175–186.
47. Zhou, J.; Mourelatos, Z.P.; Ellis, C. Design under Uncertainty Using a Combination of Evidence Theory and a Bayesian Approach. *SAE Int. J. Mater. Manuf.* **2009**, *1*, 122–135. [CrossRef]
48. Rupp, A.A.; Dey, D.K.; Zumbo, B.D. To Bayes or Not to Bayes, from Whether to When: Applications of Bayesian Methodology to Modeling. *Struct. Equ. Model.* **2004**, *11*, 424–451. [CrossRef]
49. Schmidt, J.; Marques, M.R.G.; Botti, S.; Marques, M.A.L. Recent Advances and Applications of Machine Learning in Solid-State Materials Science. *npj Comput. Mater.* **2019**, *5*, 83. [CrossRef]
50. Chen, L.; Pilania, G.; Batra, R.; Huan, T.D.; Kim, C.; Kuenneth, C.; Ramprasad, R. Polymer Informatics: Current Status and Critical next Steps. *Mater. Sci. Eng. R Rep.* **2021**, *144*, 100595. [CrossRef]
51. Medasani, B.; Gamst, A.; Ding, H.; Chen, W.; Persson, K.A.; Asta, M.; Canning, A.; Haranczyk, M. Predicting Defect Behavior in B2 Intermetallics by Merging Ab Initio Modeling and Machine Learning. *npj Comput. Mater.* **2016**, *2*, 1. [CrossRef]
52. Pilania, G.; Iverson, C.N.; Lookman, T.; Marrone, B.L. Machine-Learning-Based Predictive Modeling of Glass Transition Temperatures: A Case of Polyhydroxyalkanoate Homopolymers and Copolymers. *J. Chem. Inf. Model.* **2019**, *59*, 5013–5025. [CrossRef] [PubMed]
53. De Jong, M.; Chen, W.; Notestine, R.; Persson, K.; Ceder, G.; Jain, A.; Asta, M.; Gamst, A. A Statistical Learning Framework for Materials Science: Application to Elastic Moduli of k-Nary Inorganic Polycrystalline Compounds. *Sci. Rep.* **2016**, *6*, 34256. [CrossRef] [PubMed]
54. Seko, A.; Togo, A.; Hayashi, H.; Tsuda, K.; Chaput, L.; Tanaka, I. Prediction of Low-Thermal-Conductivity Compounds with First-Principles Anharmonic Lattice-Dynamics Calculations and Bayesian Optimization. *Phys. Rev. Lett.* **2015**, *115*, 205901. [CrossRef] [PubMed]
55. Weng, B.; Song, Z.; Zhu, R.; Yan, Q.; Sun, Q.; Grice, C.G.; Yan, Y.; Yin, W.-J. Simple Descriptor Derived from Symbolic Regression Accelerating the Discovery of New Perovskite Catalysts. *Nat. Commun.* **2020**, *11*, 3513. [CrossRef] [PubMed]
56. Walker, K.W.; Jiang, Z. Application of Adaptive Boosting (AdaBoost) in Demand-Driven Acquisition (DDA) Prediction: A Machine-Learning Approach. *J. Acad. Librariansh.* **2019**, *45*, 203–212. [CrossRef]
57. Nie, P.; Roccotelli, M.; Fanti, M.P.; Ming, Z.; Li, Z. Prediction of Home Energy Consumption Based on Gradient Boosting Regression Tree. *Energy Rep.* **2021**, *7*, 1246–1255. [CrossRef]
58. Cunningham, P.; Delany, S.J. K-Nearest Neighbour Classifiers-a Tutorial. *ACM Comput. Surv.* **2021**, *54*, 1–25. [CrossRef]
59. Bansal, M.; Goyal, A.; Choudhary, A. A Comparative Analysis of K-Nearest Neighbor, Genetic, Support Vector Machine, Decision Tree, and Long Short Term Memory Algorithms in Machine Learning. *Decis. Anal. J.* **2022**, *3*, 100071. [CrossRef]
60. Greenacre, M.; Groenen, P.J.F.; Hastie, T.; d'Enza, A.I.; Markos, A.; Tuzhilina, E. Principal Component Analysis. *Nat. Rev. Methods Prim.* **2022**, *2*, 100. [CrossRef]
61. Plevris, V.; Solorzano, G.; Bakas, N.P.; Ben Seghier, M.E.A. Investigation of Performance Metrics in Regression Analysis and Machine Learning-Based Prediction Models. In Proceedings of the 8th European Congress on Computational Methods in Applied Sciences and Engineering, Oslo, Norway, 5–9 June 2022.

62. Rainio, O.; Teuho, J.; Klén, R. Evaluation Metrics and Statistical Tests for Machine Learning. *Sci. Rep.* **2024**, *14*, 6086. [CrossRef] [PubMed]
63. Du, Z. The Relationship between Prediction Accuracy and Correlation Coefficient. *Sol. Phys.* **2011**, *270*, 407–416. [CrossRef]
64. Pool, S.; Vis, M.; Seibert, J. Evaluating Model Performance: Towards a Non-Parametric Variant of the Kling-Gupta Efficiency. *Hydrol. Sci. J.* **2018**, *63*, 1941–1953. [CrossRef]
65. Sun, Z.; Wang, G.; Li, P.; Wang, H.; Zhang, M.; Liang, X. An Improved Random Forest Based on the Classification Accuracy and Correlation Measurement of Decision Trees. *Expert Syst. Appl.* **2024**, *237*, 121549. [CrossRef]
66. Abdusalamov, R.; Pandit, P.; Milow, B.; Itskov, M.; Rege, A. Machine Learning-Based Structure–Property Predictions in Silica Aerogels. *Soft Matter* **2021**, *17*, 7350–7358. [CrossRef] [PubMed]
67. Meakin, P. Diffusion-Limited Aggregation in Three Dimensions: Results from a New Cluster-Cluster Aggregation Model. *J. Colloid Interface Sci.* **1984**, *102*, 491–504. [CrossRef]
68. Scherdel, C.; Miller, E.; Reichenauer, G.; Schmitt, J. Advances in the Development of Sol-Gel Materials Combining Small-Angle X-Ray Scattering (SAXS) and Machine Learning (ML). *Processes* **2021**, *9*, 672. [CrossRef]
69. Walker, R.C.; Hyer, A.P.; Guo, H.; Ferri, J.K. Silica Aerogel Synthesis/Process–Property Predictions by Machine Learning. *Chem. Mater.* **2023**, *35*, 4897–4910. [CrossRef]
70. Chen, Y.X.; Klima, K.M.; Brouwers, H.J.H.; Yu, Q. Effect of Silica Aerogel on Thermal Insulation and Acoustic Absorption of Geopolymer Foam Composites: The Role of Aerogel Particle Size. *Compos. Part B Eng.* **2022**, *242*, 110048. [CrossRef]
71. Chaouk, H.; Obeid, E.; Halwani, J.; Arayro, J.; Mezher, R.; Mouhtady, O.; Gazo-Hanna, E.; Amine, S.; Younes, K. Machine Learning Techniques to Analyze the Influence of Silica on the Physico-Chemical Properties of Aerogels. *Gels* **2024**, *10*, 554. [CrossRef] [PubMed]
72. Sheikholeslami, M.; Farshad, S.A.; Gerdroodbary, M.B.; Alavi, A.H. Impact of New Multiple Twisted Tapes on Treatment of Solar Heat Exchanger. *Eur. Phys. J. Plus* **2022**, *137*, 86. [CrossRef]
73. Sheikholeslami, M.; Farshad, S.A. Investigation of Solar Collector System with Turbulator Considering Hybrid Nanoparticles. *Renew. Energy* **2021**, *171*, 1128–1158. [CrossRef]
74. Said, Z.; Cakmak, N.K.; Sharma, P.; Sundar, L.S.; Inayat, A.; Keklikcioglu, O.; Li, C. Synthesis, Stability, Density, Viscosity of Ethylene Glycol-Based Ternary Hybrid Nanofluids: Experimental Investigations and Model-Prediction Using Modern Machine Learning Techniques. *Powder Technol.* **2022**, *400*, 117190. [CrossRef]
75. Kanti, P.K.; Sharma, P.; Sharma, K.V.; Maiya, M.P. The Effect of PH on Stability and Thermal Performance of Graphene Oxide and Copper Oxide Hybrid Nanofluids for Heat Transfer Applications: Application of Novel Machine Learning Technique. *J. Energy Chem.* **2023**, *82*, 359–374. [CrossRef]
76. Kanti, P.K.; Sharma, P.; Maiya, M.P.; Sharma, K.V. The Stability and Thermophysical Properties of Al2O3-Graphene Oxide Hybrid Nanofluids for Solar Energy Applications: Application of Robust Autoregressive Modern Machine Learning Technique. *Sol. Energy Mater. Sol. Cells* **2023**, *253*, 112207. [CrossRef]
77. Hashjin, R.R.; Ranjbar, Z.; Yari, H.; Momen, G. Tuning up Sol-Gel Process to Achieve Highly Durable Superhydrophobic Coating. *Surf. Interfaces* **2022**, *33*, 102282. [CrossRef]
78. Manoharan, K.; Anwar, M.T.; Bhattacharya, S. Development of Hydrophobic Paper Substrates Using Silane and Sol–Gel Based Processes and Deriving the Best Coating Technique Using Machine Learning Strategies. *Sci. Rep.* **2021**, *11*, 11352. [CrossRef] [PubMed]
79. Wang, X.; Xing, W.; Tang, G. Advances in Novel Flame-Retardant Technologies for Fire-Safe Polymeric Materials. *Molecules* **2024**, *29*, 573. [CrossRef] [PubMed]
80. Huo, S.; Guo, Y.; Yang, Q.; Wang, H.; Song, P. Two-Dimensional Nanomaterials for Flame Retardant Polymer Composites: A Mini Review. *Adv. Nanocomposites* **2024**, *1*, 240–247. [CrossRef]
81. Branda, F.; Passaro, J.; Pauer, R.; Gaan, S.; Bifulco, A. Solvent-Free One-Pot Synthesis of Epoxy Nanocomposites Containing Mg (OH) 2 Nanocrystal–Nanoparticle Formation Mechanism. *Langmuir* **2022**, *38*, 5795–5802. [CrossRef] [PubMed]
82. Bifulco, A.; Casciello, A.; Imparato, C.; Forte, S.; Gaan, S.; Aronne, A.; Malucelli, G. A Machine Learning Tool for Future Prediction of Heat Release Capacity of In-Situ Flame Retardant Hybrid Mg (OH) 2-Epoxy Nanocomposites. *Polym. Test.* **2023**, *127*, 108175. [CrossRef]
83. Kiranyaz, S.; Avci, O.; Abdeljaber, O.; Ince, T.; Gabbouj, M.; Inman, D.J. 1D Convolutional Neural Networks and Applications: A Survey. *Mech. Syst. Signal Process.* **2021**, *151*, 107398. [CrossRef]
84. Ma, P.; Jia, X.; Xu, W.; He, Y.; Tarwa, K.; Alharbi, M.O.; Wei, C.-I.; Wang, Q. Enhancing Salmon Freshness Monitoring with Sol-Gel Cellulose Nanocrystal Colorimetric Paper Sensors and Deep Learning Methods. *Food Biosci.* **2023**, *56*, 103313. [CrossRef]
85. Bifulco, A.; Climaco, I.; Casciello, A.; Passaro, J.; Battegazzore, D.; Nebbioso, V.; Russo, P.; Imparato, C.; Aronne, A.; Malucelli, G. Prediction and Validation of Fire Parameters for a Self-Extinguishing and Smoke Suppressant Electrospun PVP-Based Multilayer Material through Machine Learning Models. *J. Mater. Sci.* **2025**, *60*, 1019–1040. [CrossRef]

86. Moreno, J.J.M.; Pol, A.P.; Abad, A.S.; Blasco, B.C. Using the R-MAPE Index as a Resistant Measure of Forecast Accuracy. *Psicothema* **2013**, *25*, 500–506. [CrossRef] [PubMed]
87. Lin, Y.-C.; Young, G.; Iqbal, H.; Kondarage, A.I.; Karunaratne, A.; Jones, J.R.; Tallia, F. Effect of Process Variables on Optimisation of Mechanical Properties of Silica/Poly (Tetrahydrofuran) Sol-Gel Hybrids. *J. Sol-Gel Sci. Technol.* **2025**. [CrossRef]
88. Berg, S.; Kutra, D.; Kroeger, T.; Straehle, C.N.; Kausler, B.X.; Haubold, C.; Schiegg, M.; Ales, J.; Beier, T.; Rudy, M. Ilastik: Interactive Machine Learning for (Bio) Image Analysis. *Nat. Methods* **2019**, *16*, 1226–1232. [CrossRef] [PubMed]
89. Liang, D.; Zhou, Y.; Liu, X.; Zhou, Q.; Huang, B.; Zhang, E.; Chen, Q.; Shen, J. Wettability and Corrosion Performance of Arc-Sprayed Fe-Based Amorphous Coatings. *Surf. Coat. Technol.* **2022**, *433*, 128129. [CrossRef]
90. Misyura, S.Y.; Morozov, V.S.; Orlova, E.G.; Andryushchenko, V.A. Wetting, Droplet Evaporation and Corrosion Behavior of Various Composite and Textured Materials. *Ceram. Int.* **2024**, *50*, 19269–19282. [CrossRef]
91. Akbarzadeh, S.; Akbarzadeh, K.; Ramezanzadeh, M.; Naderi, R.; Mahdavian, M.; Olivier, M.-G. Corrosion Resistance Enhancement of a Sol-Gel Coating by Incorporation of Modified Carbon Nanotubes: Artificial Neural Network (ANN) Modeling and Experimental Explorations. *Prog. Org. Coat.* **2023**, *174*, 107296. [CrossRef]
92. Wang, Y.; Liu, K.; Zhang, M.; Xu, T.; Du, H.; Pang, B.; Si, C. Sustainable Polysaccharide-Based Materials for Intelligent Packaging. *Carbohydr. Polym.* **2023**, *313*, 120851. [CrossRef] [PubMed]
93. Zalke, J.B.; Bhaiyya, M.L.; Jain, P.A.; Sakharkar, D.N.; Kalambe, J.; Narkhede, N.P.; Thakre, M.B.; Rotake, D.R.; Kulkarni, M.B.; Singh, S.G. A Machine Learning Assisted Non-Enzymatic Electrochemical Biosensor to Detect Urea Based on Multi-Walled Carbon Nanotube Functionalized with Copper Oxide Micro-Flowers. *Biosensors* **2024**, *14*, 504. [CrossRef] [PubMed]
94. Adesibikan, A.A.; Emmanuel, S.S.; Nafiu, S.A.; Tachia, M.J.; Iwuozor, K.O.; Emenike, E.C.; Adeniyi, A.G. A Review on Sustainable Photocatalytic Degradation of Agro-Organochlorine and Organophosphorus Water Pollutants Using Biogenic Iron and Iron Oxide-Based Nanoarchitecture Materials. *Desalin. Water Treat.* **2024**, *320*, 100591. [CrossRef]
95. Kistan, A.; Mohan, S.; Mahalakshmi, S.; Jayanthi, A.; Ramya, A.J.; Karthik, P.S. Sol-Gel Technique, Characterization and Photocatalytic Degradation Activity of Manganese Doped ZnO Nanoparticles. *Main Gr. Chem.* **2024**, *23*, 17–30. [CrossRef]
96. Chelghoum, H.; Nasrallah, N.; Tahraoui, H.; Seleiman, M.F.; Bouhenna, M.M.; Belmeskine, H.; Zamouche, M.; Djema, S.; Zhang, J.; Mendil, A. Eco-Friendly Synthesis of ZnO Nanoparticles for Quinoline Dye Photodegradation and Antibacterial Applications Using Advanced Machine Learning Models. *Catalysts* **2024**, *14*, 831. [CrossRef]
97. Ricciarelli, D.; Kaiser, W.; Mosconi, E.; Wiktor, J.; Ashraf, M.W.; Malavasi, L.; Ambrosio, F.; De Angelis, F. Reaction Mechanism of Photocatalytic Hydrogen Production at Water/Tin Halide Perovskite Interfaces. *ACS Energy Lett.* **2022**, *7*, 1308–1315. [CrossRef]
98. Koyale, P.A.; Mulik, S.V.; Gunjakar, J.L.; Dongale, T.D.; Koli, V.B.; Mullani, N.B.; Sutar, S.S.; Kapdi, Y.G.; Soni, S.S.; Delekar, S.D. Synergistic Enhancement of Water-Splitting Performance Using Mof-Derived Ceria-Modified g-C3N4 Nanocomposites: Synthesis, Performance Evaluation, and Stability Prediction with Machine Learning. *Langmuir* **2024**, *40*, 13657–13668. [CrossRef] [PubMed]
99. Seitz, P.; Scherdel, C.; Reichenauer, G.; Schmitt, J. Machine Learning in the Development of Si-Based Anodes Using Small-Angle X-Ray Scattering for Structural Property Analysis. *Comput. Mater. Sci.* **2023**, *218*, 111984. [CrossRef]
100. Mu, S.; Shen, W.; Lv, D.; Song, W.; Tan, R. Inkjet-Printed MOS-Based MEMS Sensor Array Combined with One-Dimensional Convolutional Neural Network Algorithm for Identifying Indoor Harmful Gases. *Sens. Actuators A Phys.* **2024**, *369*, 115210. [CrossRef]

Disclaimer/Publisher's Note: The statements, opinions and data contained in all publications are solely those of the individual author(s) and contributor(s) and not of MDPI and/or the editor(s). MDPI and/or the editor(s) disclaim responsibility for any injury to people or property resulting from any ideas, methods, instructions or products referred to in the content.

MDPI AG
Grosspeteranlage 5
4052 Basel
Switzerland
Tel.: +41 61 683 77 34

Molecules Editorial Office
E-mail: molecules@mdpi.com
www.mdpi.com/journal/molecules

Disclaimer/Publisher's Note: The title and front matter of this reprint are at the discretion of the Guest Editors. The publisher is not responsible for their content or any associated concerns. The statements, opinions and data contained in all individual articles are solely those of the individual Editors and contributors and not of MDPI. MDPI disclaims responsibility for any injury to people or property resulting from any ideas, methods, instructions or products referred to in the content.

www.ingramcontent.com/pod-product-compliance
Lightning Source LLC
Chambersburg PA
CBHW062040220926
51850CB00010B/2612